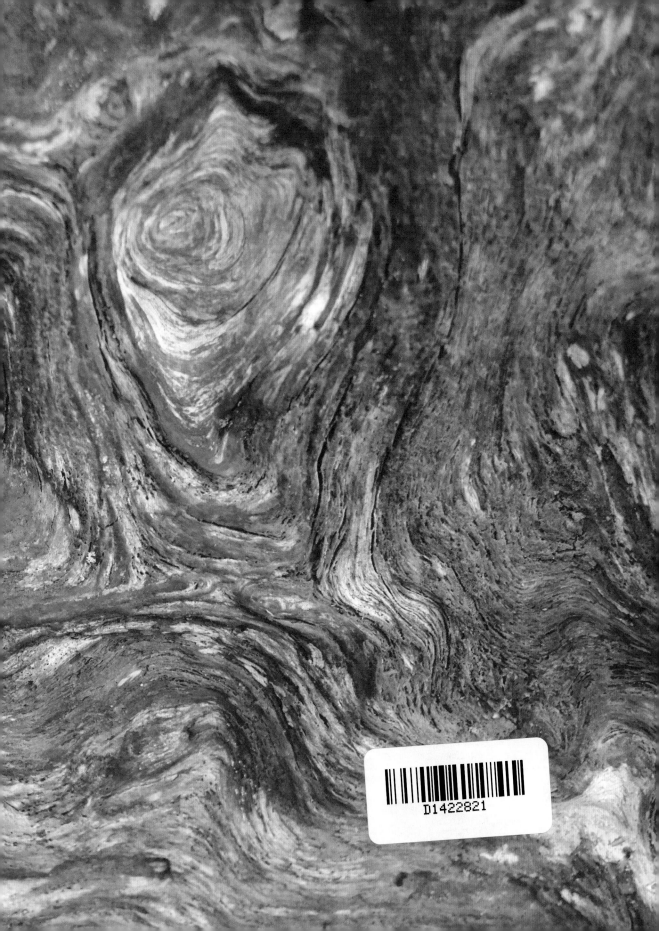

FOR THE LOVE OF
TREES

A CELEBRATION OF
PEOPLE AND TREES

VICKY ALLAN
ANNA DEACON

BLACK & WHITE PUBLISHING

First published 2020
by Black & White Publishing Ltd
Nautical House, 104 Commercial Street, Edinburgh, EH6 6NF

1 3 5 7 9 10 8 6 4 2   20 21 22 23

ISBN: 978 1 78530 309 8

A CIP catalogue record for this book is available from the British Library.

Layout by www.creativelink.tv
Printed and bound in Croatia by Grafički Zavod Hrvatske

DISCLAIMER

Spending time in the woods can be a joy but, like many outdoor activities, it can carry the risk of injury and death. Before undertaking any activity described in this book, such as wild foraging or tree climbing, please assess the risks carefully and, if in doubt, don't do it. This book is not a guide to woodland safety and therefore neither the authors nor the publisher can accept any responsibility for damage of any kind, to property or persons, that occurs either directly or indirectly from the use of this book or from any tree-based activity.

To our mums and dads who planted the seeds:
Helen and Addy, Sylvia and Stuart

There is a face in the trees—

I lost a language
to the gap-toothed birch.

Even the pine has learned how to swoon
when the wind
deposits a secret.

Alycia Pirmohamed, 'My Body Is A Forest'

CONTENTS

LISTENING TO THE FOREST, VICKY ALLAN 01

THE LONG EXHALE, ANNA DEACON 02

1 **US AND THEM**
TREES AS PART OF WHO WE ARE 05

2 **THE TREE THERAPIST**
THE WOODLAND MIND FIX 25

3 **THE GRAINS OF TIME**
THINKING ON A DIFFERENT SCALE 49

4 **ASH TO ASHES**
TREE COMFORTS IN GRIEF.. 63

5 **CARRY ON LIKE A TREE**
HOW TREES KEEP US GOING 85

6 **THIS TREE SHALL NOT BE MOVED**
THOSE THAT FIGHT 101

7 **PINING AND LOSING**
WHEN A TREE IS GONE 119

8 **TREE SPIRIT**
THE SACRED GROVE AND THE FAIRY KINGDOM..............129

9 CALL OF THE WILD
WHY JUST PLANTING ISN'T DOING IT 141

10 THE GIFTS OF THE TREES
THE ART OF TAKING ONLY WHAT WE NEED 159

11 THE DEEP, DARK WOODS
FINDING OURSELVES IN THE FOREST UNDERWORLD 175

12 THE HUNDRED ACRE WOOD
ECO KIDS AND FOREST SCHOOLS 193

13 THE CONCRETE FOREST
AN URBAN TREE LOVE AFFAIR 211

14 A WONDROUS WILD WEB
THE ROMANCE OF CONNECTION 227

15 WHAT CAN WE DO?
HOW TO FOREST THE FUTURE 241

THANK YOU 249

FURTHER READING 250

LISTENING TO THE FOREST

The company of trees is different from that of any other living being. Of course it is. All life forms have their own distinctive presence. But trees seem to have a shape and form and *there*ness that demands we connect with them. Trees were preoccupying me, my sense of their overwhelming importance, when I started to have the conversations that formed the roots of this book.

In the autumn of 2019, at a time when we were publishing a book about the joys of wild swimming, I found myself being irresistibly drawn to things tree. When my friend Karen's father passed away, I chose a card printed with an image of tree rings. If there was a place that *For the Love of Trees* took hold, it was probably there. I mentioned Richard Powers' *The Overstory* in the message I wrote to Karen, and how much it was affecting me. She wrote back, asking if I realised she was obsessed with trees.

It starts, of course, early in our life – the connection with trees, the scrunching of leaves in our hands, the gazing up from our buggies and blankets into their network of branches and leaves, and also the stories, the tales of deep dark woods and the small children who get lost in them. Then it keeps on going. Even when we don't notice them, even when we have gone tree blind, they are still there, as are the stories. The tree that we climbed when we were kids. The tree that our own child fell out of. The trees that we were shocked to find had been chopped down and left as only scabs in a nearby garden or park.

Trees can be markers in our lives. They were there, for instance, when I emerged from hospital after the birth of my first son. As we drove back through Holyrood Park, the cherry blossom seemed so searingly bright, I felt I had been born again myself into a world that was now in Technicolor.

Cherry trees – in a different part of town – were there too during the Covid-19 lockdown, when I would look, daily, out of my window at the bare, budding branches outside, waiting for it, willing it to blossom and unfurl. I wasn't alone. Across the UK there was a gentle epidemic of tree-watching. People, either awaiting that moment of bloom, or marvelling at its sudden arrival.

Then there are the other bigger stories of global significance about trees. Recent years have brought more and more of these as we've come to realise how vital trees are, not just to our local environment, but to the planet. The story of the mycorrhizal fungi that connect trees in a forest, allowing them to communicate with and support one another. The story, beguiling in its simplicity, that what we must do to save our planet is plant trillions of trees.

Naturally, I mentioned trees to my friend and collaborator, Anna as we were working together on our wild swimming book, *Taking The Plunge*. She responded with a flurry of stories about childhood trees. It felt to me like we had swum down through the kelp forests and ended up in a tall forest of landbound trees. They explore similar territories these two books – they are about what happens when we connect with the natural world, the world that is not human made, when we go to the perimeters of the concrete and brick where the water washes up, or to the cracks in the pavement where green life pushes through.

For the Love of Trees began as a journey of meeting people in trees and forests. We sat

under chestnuts, maples and pines, wandered through beech groves, waded through snow and bog. We were collecting stories and pictures, like foragers gathering conkers, acorns, pine cones, knotted sticks, delicate first blossom or beautiful autumn leaves.

But then the virus came. Lockdown shut us into our city, the pine forests became more distant, the meetings became impossible, and we were left foraging for our stories by telephone, and scouring local streets and parks for photo locations. I pined for the forest. Sometimes I felt like the people themselves were voices linking us to trees out there in other parts of the country. They were taking me on a branching path, teaching us something about our interconnected web of thought and experience, and how it relates to trees. They were putting into words feelings I had never been able to verbalise. We were gathering a small forest of voices that told a bigger, more expansive story. It told why trees matter to us, why they matter to the life living in them, around them, connected to them, and why they matter to the planet. Every story seemed to branch out to another, to some new and different person, to more and more new trees.

Vicky, author, Edinburgh, June 2020

THE LONG EXHALE

There is something about trees.

The earthy, fresh smell of a pine forest after the rain reminds me so completely of childhood, the feel of my feet squishing through the mossy ground, running fingers through the wet bracken, hearing the rustle of unseen creatures high above, the slight edginess of entering a dark forest and what might be lurking within adding to the excitement.

Or climbing the old ash tree at the bottom of our garden, numerous trips up and down a day to bring blankets, books, snacks, binoculars. Once high up in the branches, we entered a different world, a world of make believe and magic, of fascination and nature, unseen by those below, nestling into the branches in our little cocoon and watching the world outside go by.

While creating this book, I have taken more time than usual to go out and just wander. Looking up into the tops of the trees watching the birds flit from one to another, seeing the sway of the high branches, the sky behind an exquisite lace pattern. Getting down on to the forest floor to study the intricate root systems, moss, mushrooms, small creeping creatures that dwell at the base of trees. Then the bark, with its patterns and texture different for every tree, like a fingerprint. The leaves, with such a huge variety of green tones and textures, spiky, curly, soft, feathery, shiny, tough and fragile.

Spring brings the wonder of new beginnings, the tender unfurling of new leaves, the sap rising, bursts of blousy blossom, trees rich with the sound of birdsong and the smell of the earth after rain, the gradual greening of the landscape is like a long exhale after a time of holding your breath.

Summer days spent leaning back against a warm tree trunk for dappled shade, lying in the grass gazing up at the leaves, listening to the buzz of busy insects, the smell of flowers in the air. The earth feels strong, at its peak somehow.

Then autumn, the rich-toned leaves catching the first frost, the soft, low sun piercing through golden canopies, beds of rustly leaves to crunch through, spiky chestnuts, warm, smooth brown conkers, acorn cups. With layers upon layers of colour, the trees give a final magnificent display before closing down for winter.

Winter trees, punctuating the landscape with their spiky branches, their shape laid bare, almost vulnerable for their lack of leaves, but yet steady through storms, snow, wild wind and frost. I sometimes think these are my favourite trees, their roots solid despite being wildly blown around, frozen and naked, they wait for better times, growing stronger each year. They teach us a lot about life, about holding firm when our existence feels unimaginably hard, keeping faith that spring will always follow.

Anna, photographer, Edinburgh, June 2020

1

US AND THEM

TREES AS PART OF WHO WE ARE

Sycamore. Mountain Ash. Beech. Birch. Oak.
In the middle of the forest the trees stood.
And the beech knew the birch was there.
And the mountain ash breathed the same air
as the sycamore, and everywhere
the wind blew, the trees understood each other

JACKIE KAY, 'The World of Trees'

When we meet a tree, or even a whole forest, we are coming in contact with a being or beings different from ourselves. The tree has no eyes with which to watch us, no mouth with which to speak, no arms to wrap around us, yet many of us are moved. We have the sense of being in the presence of something; "something" bigger than just that singular tree. Science, even, is increasingly telling us that is the case. It shows us how trees in a forest communicate with each other through the release of chemicals into the air, or via the mycorrhizal fungal networks through which they supply food to each other. The presence of trees reaches out into the soil, the air, the thrumming life beyond.

This book is the story of a love affair – one shared in myriad ways by the many people whose stories are found in these pages. It's also an attempt, through a broad-leafed forest of human voices, to tell the wider story of why trees matter so much, not just to us, but to everything – to the vast global ecosystem that is our planet, a mind-boggling network, the connectivity of which most of us can barely get our heads round.

My friend, Karen, once described her childhood memories of walking through fiery maples in an Ontario fall, of her desire to take those leaves, some bit of the tree home. She discovered the leaves eventually would crinkle and fade, but she carried on gathering them all the same. "It's basically love, isn't it?" she said. "That's what it is."

Yes, a lot of it is love – and it's also partly, perhaps, an acknowledgement of need and dependency. Trees create what we need to breathe, to oxygenate the air that enters our lungs. There's no doubting that trees are part of our history, and part of our present. Our futures are at this current point intertwined. We now talk about trees as our saviours, the sticking-plaster to the wound that is our global carbon emissions. Yes, we chop them down, burn them, exploit their timber, but we also revere them, embrace them, craft them into objects of beauty and utility and look to them for rescue from what damage we've done.

This feeling we have around trees prompts so many questions, questions that arose again and again as I interviewed people for this book. Are we biologically evolved to live among them? Do trees have such resonance because they are part of our evolutionary history?

The answer, of course, is a bit of both. We have some primal part of us that is arboreal, bequeathed by our tree-inhabiting ape ancestors, but we are also the descendants of the bipedal, upright-walking hominids of the savannah. We have the forest in us, but also the plains. We are complicated – but trees are undoubtedly an essential part of who we are.

As Lucy Jones, author of *Losing Eden*, points out, there may even be a shape of tree, we are programmed, through evolution, to be drawn to. "It turns out," she says, "that studies suggest we still prefer trees that are the same kind of shape as those that would have helped us in our evolutionary history. An example of those trees is the Acacia tortillis, which has a broad canopy and spreads in width further than its height and would have been an ideal place for us to hide or shelter. Studies by biologist Gordon Orians found that people still prefer those shapes of trees to this day."

PETER WOHLLEBEN

Forester and author of the international bestseller
The Hidden Life of Trees

One of my big revelations came through this old stump in the forest which I had overlooked, but, when I inspected it, turned out to be from a tree felled around five hundred years ago. The stump was like a circle of mossy stones. We thinned the bark with a bit of a knife and the green layer shined up and this is always the sign of living tissue. So we saw that it is still alive, this old stump without any green leaves, just those grey stones that don't look like a tree.

That the stump was still living means it must get its energy from somewhere, and the solution was that the surrounding trees support this old stump through their root systems with sugar. Before that, I was told that trees compete for light, for space, for water, for whatever, but this made me see that trees are social and they cooperate, and that makes sense.

Some scientists now say the real tree is underground. The trunk and the branches and the leaves are the feeding organ of the tree, and the real tree, the roots and the brain-like structures are underground. Interestingly, you only find these old living stumps in unmanaged forests. In plantations and managed forests you won't find this phenomenon. In those places the trees aren't able to connect properly anymore.

JACKIE KAY
Poet

The tree that I get the most pleasure from where I live is a magnificent horse chestnut tree. My partner Denise and I often say let's go to the tree, like it was going to the pub. So we go and visit it and just stand in awe at the magnificence of it. That horse chestnut is a comfort during this pandemic. It's like running to your mammy. She has got her arms wide open.

I wrote the poem, 'The World of Trees', when I was working as the poet in residence in the Forest of Burnley. I was hooked up with these foresters and they were quite gruff at first. I think they didn't really like the idea of having a poet in their forest. They thought it was new-fangled weird thinking. But I won them round eventually, and they shared their knowledge about trees with me. Everything in that poem is true about trees.

It's true that trees are aware of each other and that they know each other's presence. It's true that if one tree leans to the east, the other will lean to the west. It's true that when a tree dies then another one that has been wanting it to die will shoot up into its space. It's also absolutely true, and there have been lots of studies that have been carried out that trees have a real sense of each other.

ALDO KANE

Adventurer

I love sleeping in the forest. I've slept hundreds of nights in different forests from high montane forests to tropical forests to swamp. If people haven't done it, I'd say it's one of the bucket list things they should do. Even better, to heighten senses, is to do it on your own.

There's quite a primal thing about being in woods. Once you've been in the forest and you've sort of attuned or calibrated to the forest, then you come away from it feeling recharged and energised.

My dad was a scout leader, so we were effectively in the Scouts long before we were old enough to be in them. Trees have been a huge part of my life in lots of different ways from being a kid, climbing in them, using them as shelter and protection. In my military days we were using trees and forests and jungle, as we were operating in there.

I've been in primary rainforest in most continents. The first time I ever went to the jungle was in Malaysia when I was nineteen with the Royal Marines. I remember the first time I flew over the top of jungle: it was in a helicopter, doors-off, at about one thousand feet. And it's like flying over the top of broccoli. That was twenty-three years ago when Malaysia had forest. As far as I could see from one thousand feet up was just forest and mist. I've been back to the same place. I was there two years ago and the primary rainforest is almost fully disappeared and it's been transformed into a monoculture with palm oil plantations.

It's pretty devastating to see. Even in that short space of time, the climate's changed in these places. Forests are usually quite a good indicator of what's going on in the world and how healthy they are. Peninsular Malaysia is one of the saddest places to see.

The first thing that I teach people when they go to the jungle is not to be worrying about snakes and spiders and jaguars and all these other things, because you're never going to see them. The one thing that you need to watch out for is what we call deadfall, and it's the biggest killer of people in tropical forests by a long way. What happens is that the emergent tree above the canopy gets struck by lightning and it will burn out, then that dead limb or dead branch or wherever the electricity passed through collects water through the normal cycle of rain in the jungle, and they become waterlogged and heavy, and eventually fall.

My best nights have been in hammocks in the jungle. The noisescape is one of the most beautiful things. If you are, for example, up in the high canopy of one of these big tropical trees as the sun starts to creep up over the horizon then, as different types of insects start their morning chorus at different times, you hear them from miles away just sweeping straight across the forest.

Adam Towler

EILIDH MUNRO
Filmmaker

Without doubt the most spectacular tree I've ever seen is this enormous ficus in Shipetiari Native Community within the Peruvian Amazon, where I was filming *Voices on the Road*. You could easily fit a room between each buttress root.

I'd never seen a tree like this before. The documentary we were filming was ultimately about deforestation, and so it was particularly poignant for us to see a tree of this size and grandeur. I think people expect rainforests to be full of ancient giants like this, but it's not the case, and so it was pretty breathtaking when we came round a corner in the forest to stand underneath this towering ficus. It was near our basecamp, so it also became a bit of a milestone for us on the way home after a long day of filming in the forest. We'd see our old friend and know we were close to putting down our bags and playing rock, paper, scissors for who got the first shower.

By spending time in communities I gained a greater understanding of the challenges involved with living in the middle of the rainforest, such as no running water or electricity, and limited access to health care or education. We wanted people to understand that environmental destruction isn't a binary choice when you're concerned about the health and safety of your loved ones, and that achieving conservation goals in the long term means balancing complex social and environmental needs.

However, at the same time, the longer you spend in a place as incredible as Manu, the more you experience the amazing impact that trees have – both on biodiversity and on you as an individual.

Ah, the joy of sitting on a branch and watching the world below. It can make you feel like you are some woodland creature – not quite yourself, yet also more perfectly you. It can make you wonder if tree climbing is what these bodies of ours are actually designed to do. One regular climber, Laura Alcock-Ferguson told me she liked to clamber up barefoot, "so you've got real trunk on trunk contact – that feeling is very physical". Does the pleasure we find in clambering in the branches relate in some part to who, biologically, we are?

Some think so. Adam Towler, a primal trainer who likes to use trees as part of his work, and who put them to particularly good use through lockdown, observes, "We would, back in our history, have been in the trees. Somehow we've lost that. Most adults can't pull their own bodyweight up. I like what the movement guru Ido Portal says: *Can you jump? Can you invert? Can you flip?*"

PROFESSOR ALICE ROBERTS
Television presenter and biological anthropologist

I like being among trees. My kids have always been keen tree-climbers and I like climbing trees myself. It gives you a different perspective on the world, perched up there amongst the leaves. But it also reminds you how flexible your body really is. We tend to move around in quite stereotyped ways, only using a very narrow range of our body's capabilities. We make ourselves environments with flat floors to walk on, chairs to sit in, and stairs to move between levels. As adults, I think we can forget that our bodies can do much, much more.

At the end of the day, humans are apes. The taxonomic family we belong to is the Hominidae, the great apes, which include us, chimpanzees, gorillas, bonobos and orangutans. And, while human ancestors became adapted to walking and running on two legs, they retained the capability to climb as well. My PhD student, Emily Saunders, showed that humans are very efficient climbers – as efficient as other apes, in fact – but we just do it in a slightly different way. As adults, many of us may not think of ourselves as natural climbers, but then it's surprising how easy it is when you try!

There's something else that's interesting about trees and human locomotion in the deep past – walking on two legs probably started in the trees. We used to think that human ancestors "dropped out of the trees" and moved around on all fours, knuckle-walking like chimpanzees do today, and then at some point, stood up straight on two legs. But research carried out by my colleague Susannah Thorpe at the University of Birmingham suggests this is probably wrong, and it's much more likely that walking on two legs on the ground developed directly in apes that were already walking on two legs in the trees. All the great apes do that – weight-bearing on their feet in trees – orangutans being the most bipedal, apart from human beings.

While our legs have lengthened over the course of human evolution, and our feet have become adapted for weight-bearing – our arms and hands are still very much those of a tree-climbing ape. We have extremely mobile shoulders – how many other mammals can reach their hands up behind their backs? We have a range of rotation in the forearm too – called supination and pronation. The radius can rotate around the ulna – which means you can hold your hand out, palm up, and rotate it almost 180 degrees to bring it palm-down.

This means you can reach out and grab a branch at any orientation – a great adaptation for a tree-climbing ape!

CLEMENTINE EWOKOLO BURNLEY
Writer

My uncle told me only males could climb trees. My mother hissed and flapped her palm beside her cheek as if she was being pestered by a too-loud fly.

Never mind your uncle, she said.

What shall I climb? I wanted to know.

A mature mango or a cashew tree, she said. You could get a Highland terrier up one if you were both determined enough.

She hesitated and looked thoughtful.

Guava trees are easy but I think the branches are too thin to hold your weight.

But what if Uncle Jeremiah was right?

I had never seen a grown woman up a tree whereas men braced their bare feet against the trunks of coconut trees and shook the nuts on to the ground for a living. Men cut spiralling grooves in the rough exterior of rubber trees and bled them for latex milk, men cut palm trees in half with machetes and drank both sweet and strong fermented sap from the dying trunks.

I never met another human being apart from my sister in a tree. We were not lonely in the company of one green snake, a multitude of fire ants, brown bats and killer bees.

BEN MEDDER
Tree-climber and movement coach

Trees invoke a play drive. You can see it across all cultures. Children have these innate play drives for certain movements that I think were probably of evolutionary importance at some point. Nature found a way to encode a reward system into movement patterns such as running, jumping, crawling, climbing, rough and tumble play.

I didn't always climb trees. I went from being very active and playing a lot of sports at school to, in my twenties, becoming very sedentary. I had an office job and went to the gym. But I soon realised I needed to enjoy what I was doing, so I started to do things that were more meaningful for me – martial arts, parkour. I found natural movement and tree climbing. I reconnected to what I was like as a child. I wasn't an avid tree-climber as a child but I have fond memories of climbing, running, jumping. I grew up in London so it would have been more about climbing and jumping off walls.

I now coach other people in movement and some of that takes place in trees. I think of climbing trees as a dialogue I'm having with the trees. Each species of tree has its own dialect and climbing each individual tree can become a conversation. For instance, most oaks you climb a certain way. If they're mature enough they're going to be quite thick and the bark quite rough. There are not going to be many thin branches, so you have to have a certain confidence and competency. Beech is similar but different. The bark is smoother. They tend

to be a bit bigger and burlier and a bit more intimidating. If you're climbing you need to do big moves. You feel exposed.

I find when I'm climbing trees there are many amazing emotional and physical and mental benefits, but overall it's stacking many needs for me – problem solving, emotional hygiene, connection to green spaces, and climbing with others is a very social thing. When I take the kids out it means I'm spending time with my family. The more you interact with nature, the more you are likely to care about it, which is a root reason why I climb trees.

CALLUM BRAITHWAITE

Tree surgeon

I do a range of aerial tree work. One of the highlights of my year has always been cone picking in Germany. We collect them for their seed, which is used by forestry nurseries to propagate the next generation of seedlings. It's especially important, therefore, to pick from stands with proven characteristics, so we're generally working in older stands which bear far more resemblance to natural forest than the Sitka spruce plantations you often see in the UK. If you wait until the cones fall to the ground they'll already have dried up and lost their seed, so climbers are sent in to pick them.

The season really kicks off with Douglas fir in August; these can grow far in excess of forty-five metres and the cones are all in the very top. We spend up to a week in the forest at a time, and use specialist climbing techniques to get up there and work safely. Then you just need to get stuck into a rhythm of putting cones in the bag, cones in bag, cones in bag...

They're covered in soft, sticky sap which quickly makes doing anything with your hands near impossible, so to combat that we rub cooking oil on our hands. We'll pick until the light starts to die, by which point you're covered in a strange mix of oil, sap and dirt. It's tough work, a blend of old-school hard graft meets modern productivity expectations, but to be able to spend so much time in these places is incredibly special to me. You'll hear wild boar snuffling around on the forest floor, while birdsong and the occasional call of a climber is the only sound that accompanies you in the tops. It's like taking a step back in time from the mania of urban tree care.

CHRISTINE THOMSON
Activist

I know a wonderful tree, a particularly impressive sycamore, who has been my friend in need during the strange and unsettling weeks of lockdown. He has a branch that could have been designed especially for me to sit on, from where I can look out over the city to the Pentland Hills and listen to the birdsong. Sometimes I add one of my own. He is always there – strong, safe and comforting. He has celebrated with me the birth of a grandson far away, mourned the death of a friend, sympathised with my frustrations and bewilderment. A friend in need indeed.

But our physical relationship with trees isn't just about climbing. Mostly it revolves around being beneath them, in their shade, co-cooned by a forest, looking up at a canopy and breathing in their air. The woods are, for many of us, where we go to escape, to get away from the world and its busyness, to find a life that works at a slower pace, a tree pace.

I suspect that much of the feeling I have around trees was passed down to me by my mum, a keen botanist. The house she grew up in, my grandparents' house, was built by my grandfather on land surrounded by mature trees. My mum recalls, "There was the syca-more outside my bedroom window, which did make the room quite dark but I think made me feel it was my tree."

My mum now lives in a house on a suburban estate and has long made it clear she never wanted to live there. We have never stopped hearing about that feeling in all the decades my parents have lived there. "Now I wonder," she says, "whether trees or lack of trees was the problem. Still we have grown them in our small garden and, given its size, we have more trees than anyone else."

BRIANA PEGADO
Arts director

A magnolia tree sat outside of my grand-mother's house in Silver Springs, Maryland, that bloomed every year and gave off the most lemony delicious scent. Grandma would spend time in her front garden tending to her peonies, the flower of choice in our family, but it was always the magnolia tree I loved. I would spend afternoons skipping down the brick lane in front of the house under the magnolia tree while Grandma swept away its fallen leaves. I would watch the flowers bloom as they unfurled from their huge buds. The leaves were two-toned, one side a deep green and the other a matt, almost fuzzy textured golden-oak brown. They welcomed me to my grandmother's house each time I visited, which would sometimes be for a weekend or for weeks on end during the summer.

That tree symbolised many things now that I look back on it. It embodied a beauty, con-stancy, poise and majesty that was a constant in every interaction with my grandmother. We spent many summer afternoons together on her back porch listening to carpenter bees drilling holes into the wood of the porch and listening to the wind breeze through the trees. The tree would grow and bloom in cycles – a reminder of the cyclical nature of the cosmos. I also lived twenty minutes away in the throngs of a deep green forest. Under the canopy cover I grew and played deep in the nature of all things. It was later on when I spent less time at Grandma's and I had moved away from my childhood home

that I felt I lost myself when I stopped wander-ing through the trees and feeling the breeze. That tree grounded me deeply in the core of myself like much of nature does.

***

Woods have also long been associated with escape. The idea of getting away to a place in the woods, of the urban population heading out to spend leisure time in the forest has been a tradition in many countries. But here in the UK, though we have our Center Parcs and log cabin holidays, there isn't quite that culture.

In Scotland there is a small tradition of hutting, dating back to between the First and Second World Wars, when workers from in-dustrial areas would pay a small ground rent to a landowner and build a simple hut for use of family and friends. Recently, a momentum for such cabins has developed around the 1000 Huts movement – but it's nothing by compari-son with Norway, for instance, where there are half a million huts to Scotland's five hundred.

I spoke with the author and broadcaster, Lesley Riddoch, who owns a hut and has written a PhD and a book on the subject. "Everyone else at a wooded latitude has access to a hut or a cabin," she says. "People in all the other countries – the Nordics, the Russians with their dachas, the Germans, the Czechs, the Americans, the Canadians, the New Zealanders. If you have a wooded environment it's pretty natural to try to take advantage of that and have escape routes."

Briana Pegado

The chief reason for this discrepancy, she says, is the approach to land ownership and forestry in Scotland. "We have this idea of forests as lockable things. There are industrial units. They are not naturally reseeding areas of land, which is what they tend to be in especially the Nordic countries. They are owned by a jaw-droppingly small number of people compared to the land ownership pattern in all the countries where there are huts. For my research, I correlated the figures. Simply put, you get more huts when you have more owners of forestry."

LIZZIE GRAHAM

Wild horseback tour guide

For a while, in my teens, I moved to the west coast of Scotland and lived alone in the middle of nowhere in an old caravan in the woods, washing in the waterfall, foraging for food. It is a protected feeling. Sleeping there, you feel muffled – the rest of the world is so far away and you feel nestled.

THE JOY OF TREE HUGS

And tree hugging is nestled among those woodland activities that are free and enjoyable. Fully wrapping yourself around a huge, thick trunk is not just a dippy-hippy gesture. Most of us will testify that it actually does make us feel good. So much so, in fact, that during the Covid-19 quarantine period, the Icelandic Forestry Service encouraged people to hug trees to help themselves overcome their sense of isolation.

Many people also sought out arboreal quarantine hugs during the UK lockdown. Among them was director of Wild Tree Adventures, Tim Chamberlain, whose mother passed away at the start of lockdown. When I spoke to him, he had not had a hug in the whole of the lockdown period, even at his mother's funeral, and was planning on hugging all the trees in his garden. "I will probably squeeze to death the first person I'm allowed to hug," he said.

But you don't have to save your hugging for comfort in difficult times. Actor Idris Elba has described how he celebrates birthdays, New Years, anniversaries by wrapping his arms round a trunk. "I just feel," he said in a *Guardian* interview (published 24.06.2019), "a massive connection to the roots that are underneath, which are very high and wide, and to the oxygen that comes from the top. And then there's me in the middle . . . Idris Elba, tree-hugger!"

DAVID KNOTT

Curator of the Living Collection at the Royal Botanic Garden Edinburgh

I've been trying to hug a tree a day for 350 days to raise funds to save the Sequoiadendron avenue at Benmore garden. A couple of staff in the garden said, "Well, why don't you hug a tree for publicity?"

It was partly tongue in cheek, and I thought well why not? Hug a tree a day. There are 3,500 trees in the Botanics in Edinburgh and there's a story behind every one of them. Every tree that we've got is different. Some of them are wild collections, some of them are historic trees, some of them are important cultivars. Every tree has a story.

One of my favourite hugs has been one I did with a pterocarya, a hybrid wingnut, that has a face like an Ent – and I don't know who has the ugliest face, me or the tree.

Trees frequently inspire a very distinctive feeling of connection, unlike that which we feel for many other living things. We relate to a tree in a way we don't relate to a daisy, a chaffinch or a beetle. There's something in their scale and constancy that makes us attach and accord them respect. Trees have place. They don't move. They are living habitats. These are the kind of emotions and experiences that inspired *For the Love of Trees*. That intensity of connection. The feeling many people have of belonging in trees – as if they have "come home".

2

THE TREE THERAPIST

THE WOODLAND MIND FIX

I love the roaring of the wind
The calm that follows cheers the mind

JOHN CLARE, 'The Wind and Trees'

Trees make us feel better. I'm not saying they always do, or that communing with a tree or forest is going to help solve every emotional or mental ailment, but, increasingly, research is showing they can help. Recent decades have brought us news that suggests being among trees can lower blood pressure and reduce anxiety and depression. Many of us, perhaps, already knew that for ourselves. We knew the mind-clearing lift that comes from a walk in the woods. We understood how a cycle along a forest trail blew the mental cobwebs away.

Some of the benefits of time spent in forests have been scientifically researched. In Japan, where shinrin-yoku – literally meaning forest bathing – has been practised since the 1980s, studies have suggested that trees may help in a diversity of ways. As Charlotte Marriott, a consultant psychiatrist with an interest in the evidence around nature-time, says, "Studies have demonstrated benefits including improved vigour and reduced anxiety, depression and anger, and regular forest bathing may reduce the risk of psychosocial stress-related diseases."

Perhaps the most well-known of these forest health benefits is that chemicals trees release, called phytoncides, when breathed in, are linked with reduced levels of stress hormones.

But there is clearly more than chemicals at work here. Many organisations now run forest bathing sessions in the UK. Among them is The Forest Bathing Institute in Surrey, where Gary Evans and Olga Terebenina, regularly take groups out to bathe in ancient woodlands and have created what they call "Forest Bathing+", an experience that incorporates mindfulness.

"Any woodland is much better than being stuck indoors," Evans says. "But we've been looking at what is the most therapeutic forest we can use for vulnerable people and groups and have already measured the increased benefits of ancient woodland. That's because there is a lot more going on – more biodiversity, wider range of colour and textures, sounds, increased variation in bird life."

Frequently, he says, people make observations about trees that help put their own lives in perspective. "The vast majority of people love standing by old trees and imagining what they've seen over the years. There's this palpable atmosphere around the trees that are six hundred-plus years old, a peaceful, serene feeling. We have also measured the benefits to our nervous system with the University of Derby. Simply being in the presence of a large tree and focusing on it leads to increased parasympathetic nervous system activity – the body's relaxation response."

Doctors in some areas of the UK are already socially prescribing walks in the woods and nature. But the Forest Bathing Institute would like to see full prescription status for forest bathing, with it being viewed as an equivalent and alternative to, for example, blood pressure tablets.

"In Japan," says Gary, "you could be given the choice of something like blood pressure tablets or a walk in the forest because they've measured similar effects." To that end, the Institute has begun working with a network of universities to reproduce the Japanese science, here in the UK.

NATURE'S MEDICINE

MIRANDA HART

Comedian

There have been many trees in recent years that have, as it were, spoken to me. I say recent years because I think had you told me that a tree might give me meaning a couple of decades ago, I might think you loopy or passed it off as wishful thinking! I remember a couple of years ago sitting in the middle of a circle of newly planted silver birches, beautiful trees at the best of times, and this happened to be during a sunset and their silver trunks seemed to light up as the sun dipped. I remember a deep peace coming over me as I sensed that they were teaching me: "just be". There they were, putting down roots to simply just be. No pressure to be any other way than who they are. And by doing that they are giving so much. Beauty, energy, joy, health to the planet.

Last summer I rented an Airbnb in Sussex for a couple of weeks' holiday. I was so lucky because just up the road from the cottage was an old train line to walk on. The trees growing on the banks had formed an arc over the line.

It truly was a green canopy protecting you. If it rained a little then you didn't get wet. If it was too hot you could dive down under the canopy to cool off. And one day I was recovering from a nasty cold and I started to feel really dizzy and unwell, which was quite scary as I was on my own. So I chose to lie down and just be under the protection of these ancient trees. I felt so calmed and held, it was like they were giving me a hug. My breathing slowed and my strength came back and I got back safely. It was a beautiful example of how to use nature.

If you feel stressed – get thee to a woodland and lie under the green canopy. If you feel tired – look upon the strength of a proud, single tree in a field. If you feel lost – reflect on the community of a copse. If you feel alone – know that the nearest tree you can see has seen it all, many people feeling like you are before you. And if you don't have access to nature itself then studies prove that even looking at pictures change your brainwaves to induce calm. A tree a day keeps the doctor away!!

ROB McBRIDE
Tree hunter and founder of Treespect CIC

The combination of long hours, stressful work, grief and modern-day living can lead to serious health consequences. My ticking of all these boxes eventually caused me to have a classic burnout or nervous breakdown, leading to me having to give up my career as a software engineer in 2004. Leaving work for the last time in a "blue light" ambulance with a suspected heart attack is a hard lesson in the need to look after one's self better than many of us actually do. Luckily it was not a heart attack, but the following two years of mental health issues were a turning point in my life. Previous to this time, my mum and dad had both died, within two years of each other.

But, as the saying goes, "As one door closes," and so it did for me. In November 2004 I was invited by Shaun Burkey of Shropshire Council to attend a tree conference at Hawkstone Park Shropshire. I met Ted Green MBE and my life changed forever. Between Ted and Shaun I was given back confidence and motivation, and ultimately, long story short … trees saved my life.

Since then I've had the great fortune to work alongside some of the best "tree professors" – as I like to call them – in the world. I volunteered with the Woodland Trust's Ancient Tree Hunt project, recorded around 3,000 to 4,000 trees, and to this day I still record trees on to their ATI website.

I have appeared many times on television and radio, once measuring a nearly 2,000-year-old ancient yew tree for BBC Countryfile.

Presenter Julia Bradbury said to me in one film piece, "Rob, trees pretty much saved your soul," which is very true.

In 2018 I was lucky enough to be taught about shinrin-yoku by the "father of forest bathing" Dr Qing Li. It was an eye-opening experience into the actual internal, scientifically proven benefits of trees. In Japan they regard this tree therapy as a form of preventative medicine and use it as such.

I know from experience just how powerful trees can be for your physical and mental well-being. A thirty-minute walk amid the greenery can rescue a bad day for me even now. Stress levels drop, blood pressure drops, your pulse slows and a whole raft of other subtle changes can and do take place.

ALASTAIR CAMPBELL

Writer, strategist and former political aide

Earlier in 2020 I started doing Tree of the Day on social media as a bit of joke. I was amazed how it took off. I'm now, in lockdown, getting so many people sending me pictures of their favourite tree – some days it's just crazy.

I've always liked beautiful scenery, even as a kid growing up in West Yorkshire, where my dad was a vet, but I've never been that conscious of liking trees so much.

It was a combination of things that changed my view of them. Firstly, I've been becoming much more environmentally conscious. A friend, Camden Council leader Georgia Gould, gave me the David Wallace Wells book, *The Uninhabitable Earth*, for Christmas, and then started nagging me, saying, "You've got to get into climate change – it's the big thing. That's where you really should be."

Then I read *The Hidden Life of Trees* and that book really had an impact on me. I'm finding now I'm just taking pictures of trees the whole time. Fiona and I go out for walks with the dog, and that's what I do – and not randomly. I'm looking for interesting trees. I wouldn't know a larch from a beech, but I like the variety of them, how they change.

There's a dead tree on Hampstead Heath that's particularly special to me. It's one where I once had a bit of a breakdown when I was going through a really bad time. I sometimes go and sit up there by it. It's just by the pond near Kenwood House and it's where I was in 2005, when I had a really bad meltdown and started beating myself up. That was when I decided I had to get help.

The tree is dead. It's an old oak and one that was ripped up in the big storm of 1987 – you can actually see the whole end of it is up in the air. It's still amazingly beautiful, but it's dead. Because it's the exact location where I kind of went crazy, it has a hold. I sometimes go up there and just think, "Well, some things die and yet carry on giving life and other things don't die and yet feel like they're dead which was how I felt. But then there can be recovery."

I wrote a piece suggesting twenty ways to stave off depression in lockdown – and one of them was enjoy the natural world. I've always enjoyed it and always liked that sense of dramatic scenery that you get in Scotland and Yorkshire.

But now I'm amazed by less obvious things like the stunning trees outside just about every second building. What it has brought home to me is that in my urban life there's an incredible amount of natural life.

SEAN WAI KEUNG
Poet

In the recovery stage of my mental illness I lived for a while in a first-floor granny flat above a mid-terraced house. I had an uneasy relationship with the family below as they would constantly shout and argue with each other well into the night, which didn't create the most conducive environment for attempting things like self-care, gratitude lists or cognitive behavioural therapy worksheets.

Behind our building was a small garden which had been fenced around to form two separate areas – one of these ostensibly belonged to the shouting family below and the other was for me. In mine the main feature was a big apple tree which loomed over everything.

During one long June I would wake up each day and look outside and the ground would be completely littered with apples. So I would motivate myself to go out with a big bucket and scoop them up, the idea being that I would be able to give the family downstairs all the apples as a gift and then tell my social worker that I had done something positive for them, which they would then take as a sign of recovery and therefore give me fewer worksheets to complete.

Every day after picking up all the apples I would place the bucket by the back door of the family downstairs and feel a mild sense of achievement before going about the rest of my day as best as I could. However, then the night would come and the shouting below would flare up and the police would either come or not

come and I would wake up the next morning and look out my window and there would be more apples again, littered everywhere.

Eventually the apple tree became some kind of gross symbol for my recovery journey. Each day I would work to make progress and then by the next morning that progress would feel null and I would have to start over again. No matter how much self-care, how many gratitude lists or worksheets I would do, there would always be more. Yet, despite this, something kept me going. Every day I would wake up and, if I could, I would still keep trying. And most of all I would still keep hoping, that eventually I would be able to start a better chapter of my life, and that in this chapter there would be no more apples left for me to pick up.

There is a simplicity to forest bathing. Not much more is involved than just properly being there, taking in the woods with the full senses, in a mindful way.

Near our hometown of Edinburgh, Anna Neubert-Wood runs sessions as part of her outdoor experience business, Wander Women. Something she wants to get across to people is: "It's no science – it's something really simple that we used to do and don't really do anymore. But we're on a journey to getting back in touch with what actually is a tree and what we can learn from it."

Many have told us this – that they were forest bathing before they even knew the word,

Anna Neubert-Wood

before even they had heard the term mindfulness. Being present with trees is nothing new.

One of the things she does is to take groups of women into a wood and encourage them each to find a tree and sit with it. "That's something which always resonates with the women so much," she says. "What I've always wanted to bring to people was this experience of slowing down and listening in to nature, watching a tree and feeling a tree, just sitting with it. People are often a bit weirded out to begin with but then they feel so soothed and calmed and grounded from it."

The silence, she observes, is an important element of Wander Women – "walking in silence and then sitting in silence in a sit spot and listening and sensing what's around". For her, one of the most powerful things, she says, "is just to sit with a tree in really high winds and feel how it is moving – how even this strong trunk has movement".

Like Anna Deacon and myself, Anna is a keen wild sea swimmer and an advocate for the mindfulness and health benefits that immersion in both environments brings. But, she observes, the forest creates a very different feeling from the sea. She uses her native German, *beheutet*, derived from the word for "hat", and meaning protected. "In a woodland you have a shelter, something that is on top of you, protecting you, a canopy, like a hat."

Also testifying to the power of forest bathing is the author of *The Natural Health Service*, Isabel Hardman. A single session of it, she says, had such an impact that it changed how she interacts with the woods. "I thought I was very good at getting in touch with nature, but when I did a forest bathing workshop it made me realise how much I was missing out on in a sensory way."

In the depths of the forest, she experienced what she felt was "Attention Restoration Theory in action". This concept is one that was developed in the 1980s by environmental psychologists Rachel and Stephen Kaplan, and is based around the idea that the kind of gentle stimulation created by viewing natural landscapes can help us overcome the mental fatigue of our busy, often digitally distracted, lives.

In fact, when she left the forest she felt so recharged that on return she had the focus to rattle out four thousand words of her book. "I'd taken proper time," she recalls, "to refocus using the woodland and I'd reaped the benefits."

ISABEL HARDMAN

Author of *The Natural Health Service*

The forest bathing session started with us having to change our focus from near to far and switch back again. Then we started listening, and it wasn't just birdsong I heard. It was also the drip of rainwater coming through the canopy, the sound of leaves scratching one another in the breeze.

We were encouraged to touch too. I had already got into the habit of laying my hand against the bark of a tree, and there's a strange sort of relief in doing that, like you're giving your cares over to this big dinosaur. But we did more than that. We were comparing textures, feeling the different barks. And we were encouraged to take our shoes and socks off and feel the woodland with our feet. I stood on a tree stump and felt its knobbliness and slight dampness.

I now do a lot of touch and feeling whenever I go to the woods and I try to listen.

For instance, there was one day when it was just absolutely pouring it down. The woods near us have got quite a lot of subcanopy of holly and once you're under that it's quite dry, so I went and sat there and suddenly I realised that what I was hearing wasn't just the sound of the rain on the tops of the trees, but the little raindrops coming through the canopy and plummeting onto the dry floor I was sitting on. And you could not only hear that but you could also see the tiny leaf fragments flying up as they landed.

PAVITHRA ATUL SARMA

Environmental researcher and entrepreneur

There's a horse chestnut and Japanese cedar tree in Lauriston Gardens that I like to go to with my son. I remember when I first saw it, I didn't know the name of the tree and I thought, Oh my god, what is this amazing tree?

I started to go to Lauriston Gardens to do forest bathing with my son to help him get over the trauma of having undergone racial bullying at school. I'd come across an article highlighting the benefits of forest bathing, its role in Japanese culture and how it can be used to reduce anxiety, stress and help cope with trauma. I thought, we're going to try that, and for the first few months after I had pulled him out of school we spent a lot of time in Edinburgh's wooded places, on Corstorphine Hill, the Hermitage of Braid and in Lauriston Gardens.

My son would spend hours sitting in the trees, exploring and climbing. I couldn't stop marvelling at these branches, how amazingly networked they are. The forest bathing worked beautifully. It gave my son chance to process things. He would start talking about his experiences and I think he never felt he had a time limit, as if being around the trees made him feel this.

There's one particular glade in Lauriston Gardens, of sycamores and horse chestnuts. You go inside and you can lie down on the floor and look up. We spent a lot of time there. He would just sit there and he never felt under pressure, I don't think. He would spend time gazing, reflecting and realise that actually he did want to talk about it. He would always talk about something when he was hugging a tree. I find it very interesting that he would be hugging, or climbing, or hanging off a branch and that's when he would feel the urge to talk.

Six months later there was a huge change in his personality. The boy who until that point would not go to a meeting with other kids or even be around them because he was terrified and would have panic attacks when he walked into a room, chose to start meeting kids and their parents in small groups. Six months of forest bathing and he was so much more confident.

My grandfather would always say, digging your fingers into the soil will heal you . . . it is very cathartic, and Arian in many ways is like my maternal granddad. He spent a lot of time digging his fingers into the soil, potting things and hanging from trees, and it made a huge difference.

LORETTA WINDSOR-MACKENZIE
Coach and mental health for life trainer

In the University of Stirling campus, there's a very beautiful tree which was on my route back to my accommodation across the golf course near some standing stones. It's where I actually did some meditation for the first time, but I didn't know it was called meditation. I sat underneath the tree and I just listened to the rustling of the leaves, felt the breeze on my skin, tried to still my mind and focus on the sounds of nature around me.

I would do this almost every day. Then later I brought a friend with me, and we would sit there together. The trunk of the tree was quite big, we would sit in the corners. I'd sit facing the standing stones, which is where the sun would go down. Sometimes we would sort of touch fingertips and we'd use the trunk itself to support our backs. I remember the feel of that bark. It was not smooth, but it wasn't very gnarly either. We would tell each other poems and I would do this meditation in my head, but I didn't call it meditation, it was just this particular chant that I had made up. I still use it now, but not all the time. It goes like this:

"I am who I should be, everything is as it should be."

I didn't know anything about mindfulness then. There wasn't even the word mindfulness around in western culture at that time in the 1990s. Years later in the green woods near Loch Leven I added to my meditation with, "Now is here, I am here."

University was extremely stressful and confusing for me. It was the first time I experienced mental ill health, which I didn't understand at all at the time. I had what I now know was depression brought on by my parents troubled separation, leaving home for the first time to a university at the other end of the country. I was also brutally attacked in my first semester by an ex-boyfriend and nearly died. I was very fortunate to survive, but I then plunged into a severe depression on and off for about fifteen years. I still struggle sometimes today but mainly I am fine as, after a lot of hard work, I found my balance, a mild anti-depressant, healthy diet, exercise and rest.

So I was very troubled in my first semester and I was reaching out for help, but there was no one available to help me. I tried everyone, friends, family; my hometown was hundreds of miles from Scotland. Nobody I knew could help me with a mental health condition, because like me they didn't really understand it or recognise it. The only person that wanted to see me at that time was this ex-boyfriend I'd broken up with but who was desperate to get back with me. And because I was for other reasons desperate, I agreed for him to come up and see me, but of course when he arrived he said right you know I'm not helping you unless we're boyfriend and girlfriend. I played along for a bit but then I said I'm sorry, I don't really want that and he lost his temper.

He tore up my room, smashed my TV, I remember being fascinated by the bits of circuit board flying about and seeing it as if in slow motion, I couldn't move, he beat me with his fists, scratched me, pulled my hair and then strangled me. Right at the end of the attack, when I was almost suffocated, the police arrived and dragged him off me. It seemed immediately after that a counsellor arrived, and said, "Can I help you?"

But I was in a complete state of shock physically, emotionally and mentally, and the last thing I wanted was to talk about it. I just wanted it not to have happened and I didn't even want to think about it. Then the police asked me if I wanted to press charges, but at that period in time and for many years afterwards I thought that I was to blame for this attack because I felt I'd led him on inviting him up, and then I hadn't been able to give him what he wanted, so it was my fault. At that time those ideas were around and there were little or no #MeToo attitudes. I felt that I was to blame for this attack, that I was worthless and I deserved it.

Having grown up in the country, I'd always loved nature and, as I now felt so cut off from people and myself, I spent all my free time in nature, particularly the woods, at Stirling University. And that was when I found this tree, and the standing stones, and it was probably about eight months after that incident I started to connect again with myself, and it really helped me with healing the wounds to my body, my spirit, my self-esteem, self-confidence, slowly rebuilding my relationship with the world as a safe place, a nurturing place. I couldn't believe it when I went on a mindfulness retreat about two years ago and they started to tell us about these things. I thought, gosh, I've been doing this for years and this is what it's called. Mindfulness!

The tree enabled me to really be able to connect with myself and the world around me again. By having that physical relationship with the tree, the noise, the smell, its place in nature, this reaffirmed my place in nature. And so, that tree experience, that particular tree has become part of who I am.

After having children I wanted to take my way of interacting with the world and people in the world further. I decided to focus on helping people in a way that would enable them to find their purpose and their direction themselves. To help them get there, I have called my consultancy Green Tree Wellbeing based on the wonderful experience I've always had with nature and with that particular tree. It feels like my life has now come full circle. An older, wiser friend.

TREES FOR TRAUMA

The woods aren't just a place to take small anxieties or everyday lows – many testify to the fact that this is where it's best to bring life's most troubling experiences. There is, for instance, a growing interest globally in nature therapy and military PTSD.

One of the world's most pioneering projects in this area is Green Road in the United States, where forest therapy is incorporated into treatments at the Naval Support Activity Bethesda. There, in a restored forest garden, dead logs are left strewn on the ground, to provide opportunities for soldiers to connect with their feelings around those they have lost in battle. It is the brainchild of retired US navy neurologist Frederick Foote, who wanted to "isolate where nature has the most effect".

Loretta Windsor-Mackenzie

The idea, however, isn't particularly new. At Craiglockhart hospital, where First World War soldiers, including the poets Wilfred Owen and Siegfried Sassoon, were treated for shell shock, medical officer Arthur Brock believed it was important to "give Nature a chance". Brock provided a therapy, including gardening and natural history activities, that revolved around reintegrating patients into their environment and restoring the damaged links to community and place.

Among those now bringing military veterans to the woods in the UK, is JP Marriott, himself a veteran, who set up Belisama's Retreat for military veterans after talking to an old work colleague who was suffering from PTSD after his time serving in Iraq.

Though JP has PTSD himself, he says that doesn't really stem from his own experiences in the army.

"I had a difficult childhood," he says. "I was outside a lot, rain or hail. I'd be under a tree, because I just didn't want to be a home."

JP was adopted and one of the events that is at the root of his PTSD, is he says, an occasion when his adoptive parents put him in the car with a suitcase and told him he was being sent to an orphanage. They left him there for around half an hour, before coming and telling him to go back in the house again.

"From that day on," he says, "I felt like a broken toy, like I wasn't good enough." The woods are where he feels "most settled". He says, "The only things that I concentrate on down there are things that I need to get on with in the woods."

STEVE ROBINSON
Military veteran

I'm a British army veteran. I saw active duty in Bosnia and when I was there had a few too many close shaves. You see things that are not particularly pleasant. You witness the malevolent side of human nature. It gets to you. It changes your view of society forever. I got typical military PTSD. There were three specific incidents, but it is an accumulative thing. When I went to Combat Stress, they identified thirty-seven separate traumas.

I knew there was something wrong after my first tour in early 1994 but you think, Well, I've just been to war, what do you expect? And at the time it was the military culture that it just didn't exist. There was no such thing. It was a sign of weakness as a soldier. So a lot of my mates from the regiment were suffering in the same way, but very little was said.

After I was demobbed I went about a corporate career, doing some very high-pressure, very well-paid jobs, jobs like head of operations for national companies and business consultancy. I worked ridiculously long hours.

Four years ago, I had a heart attack. It was a bit of a shock. But the treatment is superb, nowadays. Having that heart attack was a defining point. I decided, "Enough of this corporate slavery. I want to do something different." I didn't want to sit in cars and traffic jams and offices all day.

After my heart attack I finally sought help and was diagnosed with PTSD, which led to me spending extended periods of time at Combat Stress at Hollybush House, Ayr. I've always been a very outdoorsy type of person and while I was there, I thought, Right I'm going to start getting out to the wilds again because I really enjoy it.

But rather than what I used to do, which was, being a typical squaddie, treating the world as your gymnasium, treating everything like a test of endurance, I started to take more notice of nature and got into bushcraft.

A key part of the treatment at Combat Stress was to educate and they helped me understand how my brain was working. I became aware that my symptoms just disappear when I'm deep in a forest. I can be really stressed, hypervigilant, borderline violent possibly and I can walk into a secluded forest, where I know there's no one within a mile of me, and I can just sit there all day and I'm totally chilled. I think it's because I know there are no people there.

So I thought that I needed to share this, and I experimented with setting up a woodland veterans retreat. Not long after that, I found out about Belisama's Retreat in Lancashire. JP Marriott, who runs it, offered me a directorship.

I love going to places like the Galloway forest where there are thousands of hectares of nothing but trees, and there's hardly any footpaths. I like exploring. I like being in charge of my own destiny. One thing I like about being out in the woods is I'm totally responsible for my own actions and safety and wellbeing. It's outside society.

MERLIN HANBURY-TENISON
Rewilder and military veteran

I grew up on a farm called Cabilla in Cornwall, where I still live. It's always been my future and my passion and the bit I have loved most is the eighty acres of ancient oak woodland we have around a gorgeous river that runs through Bodmin Moor.

I was in the military for eight years and I did three tours in Afghanistan and when I came back, like many veterans, PTSD was something I had to learn to live with. My medicine was the woodland here, the oak trees. They have become my doctors.

I properly started to appreciate the woods after the PTSD started getting really bad in 2017. At the time I was trying to weekly commute between London and Cornwall. I'd be in London, Monday through Thursday, and then in Cornwall for the weekend.

I think that almost-forced immersion and then removal, every week, can be quite traumatic in itself. It's a dramatic flip from the deep countryside which we have here into an absolute urban environment. I noticed that I was well while I was in nature, and I was unwell, both physically and psychologically, when I was in the city.

I started to develop a constant habitual cough and to feel more slow, sluggish and lethargic, when I was up in the city. That would be very quickly remedied when I was back down in Cornwall. I realised that I was making myself into a test subject showing very clearly that, with the same diet and the same activity, I was

well when I was in nature and I was ill when I was in the city.

It was about ten years after my first tour of Afghanistan that I started to receive therapy following a breakdown in 2017, precipitated by memories that were starting to come to the surface from Afghanistan. I'm lucky that my wife and family were very supportive and I didn't try to hide and self-medicate, which is a mistake a lot of people make. I sought the help that I needed and ended up being treated by an NHS organisation called the London Veterans' Service.

So many things came together to show quite how important nature is. For me there was having the breakdown and then realising I felt much happier and healthier when I was in nature, and also reading, at the same time, books like *Wilding* and George Monbiot's *Feral*, which for those of us who take an interest in this have become gospels.

It began to make sense to me that the way we're living as a civilisation is making us unwell, and a deeper connection with nature might be better for our minds and bodies.

My father, the explorer Robin Hanbury-Tenison, was very interested in this. He has always been ahead of his time, very experimental and interested in trying new things on the farm, and also a man who is incredibly in tune with the natural world.

The epiphany moment was when I was walking through the woods, feeling invigorated as a result of spending several hours in them, and at the same time thinking about some of my former military colleagues who I knew were also having their own struggle. It occurred to me that they probably didn't have access to woodland like this – so few people do – and that this shouldn't just be a place where I can heal. I want this to be a place where lots of people can heal. That was the spark that led to the idea of really turning this place into a forest bathing retreat.

THESE WOODS ARE FOR ALL

Most of us know time spent in the woods produces some kind of calming feeling. Research is starting to show that impact. A 2019 MRI study, for instance, showed that time spent in nature has an impact on the brain and how we process feelings. Psychiatrist Charlotte Marriott describes how, "green space exposure improved activity in the dorsolateral prefrontal cortex, an area of the brain responsible for processing negative emotions and stressful experiences."

What was remarkable too, she says, was that "the positive effects of nature exposure were particularly evident for the subjects who had reduced dorsolateral prefrontal cortex activity at baseline, suggesting that those people with the least capacity to self-regulate negative feelings benefit the most."

Increasingly, mental health professionals are taking such effects seriously, and even if the research is still far from substantial, there are now many groups around the UK who take people into the woods. Forestry Commission Scotland, for instance, provides a service called Branching Out which takes adults with severe and enduring mental health problems on a twelve-week programme of woodland

activity. Reviews of the work have found that participants reported small but significant improvements in mental health.

One user of the programme, designer Helena McGuinness, describes how she discovered, "As you enter the woods the noise and light levels change. There is an open ceiling of branches and leaves. There are no hard noises underfoot. There is a quality to the air you breath which is refreshing. You become aware of birdsong, or rain on the leaves or the feeling of silence. Your attention has drawn to your immediate surroundings outwards from thoughts and inwards from your normal environment. You are in the moment."

In Yorkshire, a project called Rooted For Girls runs courses taking small groups of teenage girls from diverse backgrounds out into a woodland. Its founders, Jenny Biglands and Beth Webber, say they perceive nature as "an incredible environment" for enhancing wellbeing, but one which too often teenagers, and particularly teenage girls, are disconnected from.

Many of the girls who have participated testify to that sense of wellbeing. One, for instance described, how previously, she "hated" nature, but had now discovered something else there. "It's good, it's peaceful and it relaxes your mind. Seeing the greenery and everything makes you like calm your brain and stuff."

Another talked about the way the woods made her feel relaxed and allowed her "to express my feelings the way I wanted".

Who gets to experience the impacts of time spent in green space is, of course, a big issue. One of the things Rooted For Girls is attempting to do is bring the woodland experience to those who might not otherwise go there. The issue of inequality of access to nature connection

frequently came up in conversation. The woods, people would say, too often belonged to middle-class white people, many of them landowners. Even tree hunter Rob McBride described how he felt his working-class background meant he was treated an oddity. "The tree world is very, very middle class," he said. "I've had almost shock from some people, when I talk about trees."

Among those aiming to address this inequality of nature experience is Beth Collier, who runs Wild in the City, an organisation delivering bushcraft and ecotherapy activities to people of colour. Beth sees the loss of connection to nature, in black communities, as a trauma – and one of her aims is to alter the UK situation in which BAME people are all too often alienated from, and not welcomed into, natural environments.

"Black dismissiveness of nature isn't just simple dislike," she has written in her blog. "It belies a legacy of trauma and human interference which has made being in nature feel unsafe and unwelcome. In both the US and Europe we spend less time in nature than white people. This means we benefit less from a positive relationship with the natural world and all of its health-giving properties. Many of us have been disenfranchised from nature by racism and the legacies of slavery, colonialism and poverty."

Beth is one of a new breed of psychotherapists developing pioneering nature therapy approaches which take the woodland experience much further than simple forest bathing. For her, the natural environment can be like an extra psychotherapist, part of a three-way relationship that her Nature Allied Psychotherapy approach revolves around – human therapist, client and nature itself. She tells her story here.

BETH COLLIER
Psychotherapist

I grew up in Suffolk in a very tiny village in a rural smallholding. So I was very blessed to have a childhood immersed in nature on a small farm with fields and woods beyond it. What it gave me was an uninterrupted relationship with the natural world. I came and went as I pleased. I roamed and explored and found out for myself what plants could be useful in terms of supporting me when I was out and about, what I could eat. I wasn't really aware at the time that there is a name for those sort of practices, of bushcraft.

Then as an older adult I trained and learned more about it. I think those formative experiences set me up to appreciate how there is a relationship to be had with the natural world, and emotionally it's very supportive. It was a place of feeling very much where I belonged and accepted. So I would seek out natural spaces and feel part of something.

One significant experience I remember from childhood: we had an orchard, and it felt like a real learning ground. I had so many significant moments there, and a lot of my learning about the natural world has come from a sense of communion, of being in that space, feeling that the birds and the plants are sharing things with me. I remember pairs of birds nesting and bringing food to their chicks, and it's the kind of knowledge that I couldn't articulate with words.

There was also a moment when I was about eleven years old, when I was in the orchard and

I had uninterrupted views across the forest. I was thinking how lucky I was – despite some of the challenges at home – that I had this kind of space to be in, this world in which I felt I belonged. And then I started thinking, What on earth would it be like to have these issues going on but be in an urban area, not have the freedom, green space and the beauty of green space?

The challenges I had then were of being a black child adopted by a white couple that perhaps didn't understand race or the impact of isolation in growing up in that kind of space. It was emotionally challenging at home. Nature became a place of solace and support.

I went off to university in Manchester, and I had this revelation of how much being in nature meant to me, because I fell out of sorts, and I wasn't sure why. I eventually realised it was because I hadn't been out in nature for a long time.

It brought me back to just how important that connection is. It's like any relationship. You need to keep in touch or things can drift. I now work as a psychotherapist with people who have trauma both in their relationship with nature, and in their human relationships. The journey towards that began when I was working with a young boy, about ten or eleven years old, and we were indoors, and there was nothing age-appropriate in the room. He lived in inner-city London and there was a lot of abuse and neglect at home. He was falling

into gangs. He presented as a sort of thirty-year-old, very masked and protected with this bravado, but he was a very hurt and damaged boy, underneath.

And it made me connect with my own experience of being that eleven-year-old in the orchard. I was thinking about what it would be like if his environment changed, if he had a space to just play out these angry energies.

I thought, when I know what nature does for me and others, why am I working indoors? I realised that my own experience of being in nature gave me something. I started developing my approach of Nature Allied Psychotherapy.

So now I work exclusively in natural settings, helping people explore human social relationships but also their relationships with the natural world.

Work like Beth's is a reminder that nature is for all. It's not some spa therapy for the middle classes, rewilded with a dash of mud. Nature isn't just something that we can use like a pill, or another tool we can mindlessly employ to make us feel better. It's the thing we are part of, that we evolved with. We are it. When we are disconnected from it, we are disconnected from ourselves.

3

THE GRAINS OF TIME

THINKING ON A DIFFERENT SCALE

Jemima Thewes

*The forest
keeps different time;
slow hours as long as your life,
so you feel human.*

CAROL ANN DUFFY, 'Forest'

We humans are, on average, a mere eighty rings in the life of a tree. The tree that was there when you came into this world will be there when you're gone, if all goes well for that tree. Its rings will keep on coming, cycles of growth edging outwards. This growth puts us in our place, makes us feel small, in a way that is often welcome. Trees not only outlive us as individuals, but they've been here far longer as species – an extraordinary 370 million years to our ancestors' six million – and most likely will stick around much longer after our collective demise.

Time is very often mentioned when I interview people about trees. When we look at them, we see organisms that tick to a different clock. What's fast for a tree is slow for a human being. As my friend, Karen, observed, "Trees are like community elders. They've been here longer. If you feel like quite a frenetic person, there's this sense that there's something to look at as an emblem of security or stability, that is always growing at a pace different from human life."

We pine for that slowness. In these busy times, we crave the pause that trees show to us, that they invite us into.

The very oldest of trees possess gravity, and a kind of magical improbability too. A tree hollowed out inside, half dead, half alive, can appear as an arboreal miracle. Yet, we know now that stumps can live for hundreds of years, supported by other trees, dead on the surface yet alive underneath, as if on an earthy life support.

Storyteller Amanda Edmiston talks of such trees – those that are hollowed, or with holes underneath – as having an eerie quality. They offer, she says, a liminal space. They are where the fairies dwell. "Hollow trees can live for hundreds of years," she observes. "They become almost like a portal between the living and dead."

We can feel small and ephemeral compared to such trees, like newcomers. Given our own short lifespans, a tree that can live for several thousand years seems close to immortality– like the UK's oldest tree, the Fortingall Yew, or Methuselah, the bristlecone pine in California believed to be five thousand years old.

Jemima Thewes, a forest circus performer, who plays Granny Lichenleaf, a 642-year-old character, explains how this idea of age resonates with her. "It's the fact that trees are here for so long and we are here for just this snippet and with all our arrogance think we can just chop them down."

But tree-time isn't simply linear time. It is, as druid Liz Harris described to me, seasonal and circular. It reminds us that things die back and regrow. Cycles of growth and decay, spirals and returns.

In a beachside café, Dr Coralie Mills passes

Dr Coralie Mills

over a sliver of wood. This, she tells me, didn't come straight from a tree, but was a piece of timber taken from the roof of St Giles Kirk. According to her dating, the tree it came from was felled around 560 years ago and was brought from one of the last remaining extensive medieval reserves of old growth oak in Scotland, the Royal Forest of Darnaway, in Morayshire. The rings of this oak take us back through a further three hundred years of history. It's like a snapshot from the past, which in turn tells of some other deeper past.

Coralie is a dendrochronologist: she studies the growth rings in trees. She takes such pieces of wood and makes them read like history books. "You can," she says, "find out so much. It's remarkable. It can give you the date to a year. Radiocarbon dating can't do that. Dendrochronology can tell you not only what country a tree came from, but which region. It's magical. That's why I love what I do."

When Coralie studies tree rings she is not simply uncovering the age of the trees themselves, but discovering a story about our landscape and how we've managed our woodlands, or, as she puts it, "how we have and haven't looked after them".

She recalls that when she first started to work

in Scotland she was keen to find "reserves of very old tree rings", essentially historic woods, preferably old oak woods. But she found these were disappointingly few and far between. "I've looked at a lot of what I call historic woods across Scotland and to find anything that pre-dates the 18th century is pretty hard. A lot of our historic woods come from the late 18th to 19th century, when there was suddenly an increased demand for home-produced timber because of the Napoleonic Wars, which cut off many of the foreign sources of materials."

There is, however, such a reserve at Dalkeith Country Park, where, according to studies of a number of deadwood samples, there are oaks that date back to the 16th century.

Taking samples from these dead oaks in-volved obtaining permission from Scottish Natural Heritage because the park is a Site of Special Scientific Interest conservation area owing to the presence of some very rare beetles living in the dead wood. This highlighted to Mills something about the way we value trees. "It seemed odd to me," she says, "that the reason that the oak wood is protected, desig-nated SSSI, is because of the beetles and not because of the trees – they're only indirectly protected because they are the host and habitat for the beetle. There is no universal protection for ancient woodlands and I think we humans need to wake up to that and look after what we've got left."

Woodland cover in the UK has improved since its lowest ebb just before the First World War, and now stands at 13 per cent and in-creasing. "But," as Coralie points out, "the vast majority of the cover is modern plantations. The message I would love to get out there is that we really have to look after and expand

more the tiny fraction of ancient woodland we have. They're so precious for both their natural and cultural heritage."

Coralie has interests that range beyond his-toric oaks and their almost aristocratic rings. Her biggest passion is for "working trees", her favourite being an old pollarded ash, which can be found in a stand of such trees by a Scottish loch. "I did some tree ring work on those ash trees and we found that the oldest one prob-ably went back to the 17th century. To find an ash tree with such a long lifespan is really something. And it's a massive tree that I'm thinking of. Huge. It's got this beautiful pollard structure."

The ash had been managed, she says, proba-bly to provide a mixture of firewood and poles for fencing, as well as leafy fodder for livestock over winter. Such evidence of how ordinary people were involved in woodland, is, she says, hard to find in Scotland, "because of the way land ownership has been – the laird owned most of the valuable trees and the ordinary folk had very limited access to wood".

The age of these very old trees do, she adds, make her aware of the briefness of our human lives. "It really does emphasise what short-lived creatures we humans are and also it makes me very respectful of those foresters in the past who had a view to the future beyond their own lifetime. They're the ones who did us the great-est service because we're still benefiting from the way they planted and managed woods. You often don't know who those nameless foresters were. The creation of woods often gets attrib-uted to the landowner and not to the person or people who put the acorns or the saplings in the ground, or did the coppicing. They'll be nameless."

JUDY DOWLING
Ancient tree hunter

I started recording trees for the Woodland Trust big time after I retired as a nursery teacher. I had read in a book a little bit about the Cadzow Oaks, a group of over three hundred ancient oaks in rare medieval woodland pasture. These oaks were not recorded in our inventory. So, in 2013, when I was driving up the M74 from Yorkshire, at a time when I was up and down every few weeks to see my mum, who had dementia, I decided to take a look at them. It was March and it was quite a dreich day, gloomy and dark, but I was really so fired up to see them.

Scottish Natural Heritage had given me this really convoluted list of directions for how to get in, which meant I had to go down this very long single-track road. At this point it had been snowing, and it felt like madness. I could see these trees in the gloaming, hundreds of them. I went in and as I stumbled into the middle of them, it started to snow. It was blowing a blizzard through them and there were all these looming black shapes, every one different, gnarled and wizened. I just stood there and wept. I walked among them, touching them and I thought, *Oh my God, none of these are recorded!* which was heaven to me . . . I will never forget that day.

There is also a field maple I love. When you are under that tree, it's like being enveloped in warm arms. The tree just gives me peace and a time to breathe and to sit and think, Well, this tree has been here all these hundreds of years. It has coped with all weathers and wars, and it's still here and I can be the same. I can carry on.

FOR THOSE AFTER US

Among those who are most acutely aware of the long game that is the life of a tree are those who plant or care for them. Foresters are often dismissed as in it for short-term gain, but there are many who see a forest as a project that is not about their children, but their children's children. One such is Francis Ogilvy, who has planted tens of thousands of trees at Winton, his family estate in East Lothian, to be managed using a sustainable, continuous cover approach.

FRANCIS OGILVY
Estate factor

I see myself as a caretaker. Across the generations we have been a family of tree planters, adding significantly to the area under tree cover at Winton. My dad was keen on forestry and I caught the bug early on, despite the significant annual cost. It is something we believe is a responsibility of ownership. My biggest effort has been planting one hundred acres of farmland in 2006 which equates to nearly 70,000 trees. In fact, planting was the easy bit; it's the looking after them that takes more effort – certainly for the first twenty years.

It's quite humbling when you're dealing with trees hundreds of years old; you wonder what was going on when they were planted. Rotations deal in centuries so each generation benefits from the past and has a responsibility for those to come.

I decided at seventeen I wanted to be a part of "shaping the countryside" and chose rural surveying as a profession to do this. I developed a love of trees from both sides of my family, but notably my uncle Pat. The countryside has much more to offer than I believe most people are aware of and forestry has a key part to play. I try to facilitate diversity and multi-purpose forestry. This can be countercultural to commercial practices, but it encourages resilience, habitat and opportunities for woodland use. It's an expensive gamble but hopefully will lead to cumulative benefit overall.

"There are twenty-four whole hours in every single day," said Gordon Watson, a wonderful forester friend and mentor when I trained on the Glamis Estate. In other words – what's the hurry? What indeed! Coincidentally, his father was head forester at Winton in 1903 and his legacy is part of what we see today.

STAN DUNLOP
Gardener

I'm ninety years old and, if I'm honest, when I'm with some of these trees in Starbank Park in Edinburgh I think, Well pal, I hope I can spend another year with you. Because they'll be here when I'm gone. They're there for the long stay. I always say the trees are the best watch I have for the seasons. You might have a sunny day and say, "Oh, spring's on its way," but if you watch the trees they'll not come out till the right time.

Unlike so many other living organisms, trees represent constancy. They are the unchanging marks on our well-loved landscapes, they are signatures of home. We trace our own lives against theirs, watch how they grow and change, follow the seasons. Marjorie and Gabriella here tell their stories of trees watched and returned to again and again.

MARJORIE LOTFI GILL
Poet

I've always envied those who grew up in the country, have a particular landscape almost personal to themselves, their own kind of vision. I've imagined that the view from a bedroom or a kitchen, say, could serve as a framework for seeing the world, one that you could then take with you wherever you go.

Our first home in Edinburgh looked out to Blackford Hill. And although there were other houses between ours and the hill, the children always called it "our hill". We looked to it for snow and frost, for light against the sky, and even to judge the time by the length of sun across its ridge. Now, we don't have a hill to watch, but I have a magnificent old lime tree that's in direct view from both my bedroom and the sitting room. It's become an anchor – I judge the sky against it, watch the moon move behind it on sleepless nights.

If I'm trying to think through a problem, more often than not I find I'm settled with a cup of tea looking at that tree, almost as if it's keeping my eyes busy, so my brain can get on with the work of its own.

In lockdown, it occurred to me that the tree has become that same view that I've always envied. And with that thought came the next one – not that I must never move away from it, but – that I can create the perspective I've always envied in those who are born, grow up and live in the same place, simply by finding a tree to watch.

As a girl at my ballet classes, I was taught to choose one spot to watch during pirouettes, to keep coming back to it, so I didn't get dizzy as the rest of the world blurred. For now, that lime tree is the spot I'm watching while everything else spins.

GABRIELLA ORDE
Jewellery designer

Flowers are beautiful, but so fleeting. But trees – trees have history; they can tell a story and change a landscape.

My life has been a series of unusual – sometimes unenviable, but always interesting – moves. However, a few continents and languages later, it still feels like home when I'm back in Scotland. Whenever I come back to stay with family near Inverness, one of the first things I want to see is my apple tree in the garden.

I planted that tree from an apple pit when I was in Brownies. Before I even got my first Brownie badge, before I had any idea of what life would bring, before I knew that most other countries existed, and certainly before I had any idea that I would be experiencing them. When we moved away, my little tree, which was by then less than a foot tall, went to my grandmother's. In time she moved into a flat, and so my tree made another move to my aunt and uncle's house in Scotland, finally planted in the soil that I had left behind so long ago. Visiting after many years, I found the most beautiful surprise: my tree was full of conker-sized, terribly sour, green apples! It was almost as if I was looking at a parallel, foliaged version of my own story; so far from where we both started, and yet now so sure of ourselves.

We humans project on to trees an idea of them as living witnesses. Often, in interview, people would reflect on what they imagine had gone on around a particular tree – what historic events it had seen. Trees may not have eyes, but nevertheless we talk as if they did. Rebecca Ward, who volunteered to do forestry work when her "depression and anxiety were high" recalled that the maturity of the trees was a comfort to her. "Their large stature and sense of permanence was reassuring. When I imagined what a large slice of history the great oaks had seen, it felt like they were older, wiser friends imparting years of experience."

Myrtle Simpson

Explorer Myrtle Simpson described how she found relics from the past of the forest she lives in. "This forest is old," she said. "We found items from the troops coming back from Culloden in the forest behind my house, because this was the track they took. Digging in the garden we found bits of metal bridles from the horses. We know these trees are old."

Our oldest yews speak of long histories. As yew historian and author of *The God Tree*, Janis Fry observes, yews were often *axis mundi*, centres of tribal territories, and therefore sites where key events took place. In her book, *The Ankerwycke Yew: Living Witness to the Magna Carta*, she describes how one of British history's most important political documents was sealed there. "People performed important ceremonies of state under sacred trees," she says. "In more ancient times the tree was the god, and for the tree to bear witness was like swearing to god. The tree is both the living witness and is immortal."

But, let's not forget the tree as witness to terrible deeds. Dule trees in Scotland were used as gallows for public hangings. The witches of Salem were executed on a ledge of trees. Shameful histories are attached to the lynching trees of the United States. The British filmmaker Steve McQueen, when he was looking for a tree around which to film a lynching scene in *12 Years A Slave*, came across one which had been used for exactly this purpose. Not only did he film a scene at the site, but went on to create a photographic image, titled "Lynching Tree". The shot makes for a haunting artwork – though idyllic and like many a tree image – reminding us that murder and atrocity are, part of the cultural resonance of trees. It's a chilling reminder of who we are and who we have been.

Trees connect us with our most virtuous and terrible pasts. They keep going there, keep growing and dropping leaves, as we humans live, give birth, die, fight and make peace around them. They are both our gallows and our trysting trees. Our inhumanity and our humanity.

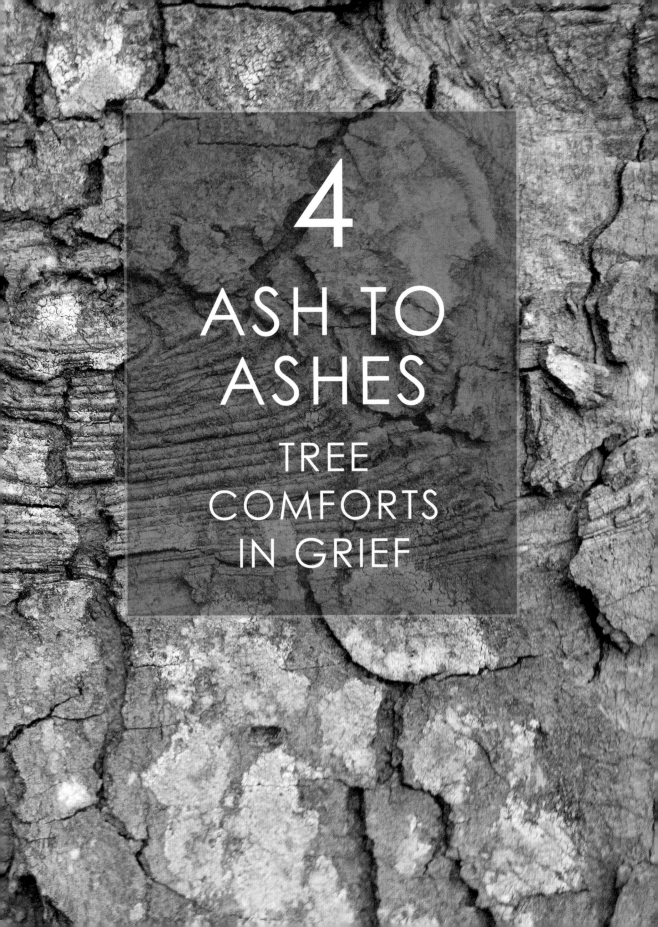

4

ASH TO ASHES

TREE COMFORTS IN GRIEF

or could it have been a representation of the tree by the burial plot he had acquired
the one on the slight slope facing eastwards
at the bend of the small foresty path
at the south end of the cemetery

the same place where i stood that day
when i said goodbye

SEAN WAI KEUNG, 'Where is the tree my gongong drew?'

Sun dapples through the bare branches of beeches. The ground is carpeted with a sprinkling of old copper leaves, many of which have already mulched down into the soil beneath. It's March and they've been sitting there over winter, breaking down. Their tenuous presence speaks of the passage of time, of decomposition, and a cycle of life going on beneath our feet. But there is something else underfoot too – something you don't notice at first, but only comes to your attention when you spot a bunch of flowers or a discrete stone marker. There are people here. This is more than a wood; it's also a graveyard.

We are in Binning Memorial Wood in East Lothian, where we have come to meet a friend, Ruth, who buried her son under these tall beeches just last autumn. It's now the run-up to coronavirus lockdown and death feels close – more present than usual. We can see Ruth there, in the distance, kneeling by the grave with a friend, rearranging the mound. She's still waiting for the stone to mark her son's unassuming spot there among the leaves.

The modesty of the grave markings is partly the result of the rules around what it is possible to leave in the wood. Sarah Gray, who, with her husband, runs Binning, says, "We've asked that people don't leave garden gnomes, paraphernalia, teddy bears, football shirts, birdboxes, just natural things."

The trees determine the seemingly random layout of graves. "We work with the trees. So everybody is at funny angles. As long as we're three metres from the centre of each tree, we're not damaging any roots."

One of the fascinating things about Binning is that the burial ground was created here partly because the Grays were keen to protect the forest. Sarah's husband had discovered that once people were buried on the land, the use of the land could never be changed. "We're farmers and we're surrounded by houses going up everywhere. Our views are changing every day and we thought, Are we going to have any of these spaces left?"

The Grays are not unique in preserving woods through the bones in their soil. In Germany, the great tree advocate, Peter Wohlleben, told me how he had created a woodland burial ground in order to protect "an old beech forest because otherwise it would have been felled".

He continues, "The community which owns it was able to get money from selling the old beech trees as living tombstones. I think it's a really good idea because it's also the best symbol for the circle of life." It's something Peter says he would like for himself. "For me I think it would be the best way, to be part of the forest. I think I will go back to nature where I come from and always have been."

It happens that Tim Maguire, the celebrant who presided over my own wedding, has visited Binning Wood and conducted woodland burials there. There are, he says, many places where you can bury a loved one and plant a tree, but, he observes, "What's special about Binning is it's a mature beechwood. The trees are about sixty years old and when you go there you think, well where's the graveyard? Then you look down."

It's there, of course, beneath the blanket of leaves, human bodies, laid to rest, their molecules becoming part of the humus, the soil, the earth, the trees. This, on some level, is where people go to become beech.

What Tim feels this natural setting speaks to is "our desire to really return to the earth and be reborn". This idea of return and rebirth is there even in the Christian line, "ashes to ashes, dust to dust", words that Maguire recalls were shunned by the humanist organisation when he was training as a celebrant. "But," he says, "the scientist Brian Cox has said, 'We are made from the same material as the stars and when we die we are returned to those materials and those in their recombined form are new life.' The idea of us going back to earth and your body making new life is the thing which I share as a humanist with many people of faith."

CHRIS PACKHAM
Naturalist and television presenter

There's a beech in the woods near where I live in the New Forest that is one of the most magnificent organisms on earth. In those woods there are oak trees which are in the region of eighty to one hundred and fifty years old, there's lots of hazel and holly, there's a very old yew tree – but none of them hold a torch to that beech pollard.

I've measured the tree and people have told me it's between 550 and 650 years old. But its age is detail. What the tree has, that all of its neighbours don't, is enormous presence. For that reason it's become a bit of a shrine. Its growth form is like a massive crown, and the basket of boughs reaches up into the sky from this enormous trunk. Sometimes you stand in front of a piece of art or a building or some natural feature in our landscape and it just puts everything else in its place. It humbles everything else. That's what this beech tree does.

At some point I will be cremated, and the idea is that we'll mix my ashes with the ashes from my last dogs, Itch and Scratch, who have been cremated, and they'll be scattered beneath that tree – and the three of us can be part of it.

Sarah Gray

Tim Maguire

I know that my carbon and theirs will mix together and collectively we will be part of that tree. In terms of romance I personally can't think of anything more romantic than that. What a thing!

Often I garden beneath the tree. I don't cut anything but I pick up and move away all of the fallen branches, even down to the twigs so that the space beneath it is clean and it stands therefore in its own arena, in its own place, on that platform of its own dead leaves.

There are two times when the tree really shines. One is at the beginning of May when it leaves. When the leaves come out, they're that beautiful verdant green and when the sun streams through them they sort of glow that yellow green and they bathe you in that green light. There's nothing else like it on the planet at all. That is my favourite time. If I go and stand there, the feel-good feeling that goes out of the tree and into me is phenomenal. It seems to encapsulate in an aesthetically outstanding way the sheer magnificence of life and new life. It really is the herald of spring.

That's the aesthetic side of it – its sheer beauty and magnificence – but then I think there's the human side of it too. The beech doesn't just humble the rest of the trees in the wood, it humbles me. You go and sit underneath it and it puts you in your place. This is something that is potentially 650 years old and I will be sat beneath it for just a fraction of its extraordinary life. I think, if trees could talk, what would it tell me about all of the love, the hate, the kisses, the tears, the passion, that it's witnessed beneath it?

A tree like this reminds us we're not the be all and end all of everything. When we sit beneath it, or even older trees like some of the yews, it

puts us in our place. There are yew trees I can access that are easily two thousand years old – their reproductive cycles, their life cycles, they have evolved to work on that sort of timespan. It's beyond our comprehension. Those yew trees were growing before we had our language, before we had our religion, before we had our countries.

Trees were important to me in my childhood, but the trees that were most important were the ones that had kestrel nests and sparrow-hawk nests in them. I took a kestrel from the wild in 1975 and I kept it and trained it. Sadly my kestrel died, and every year on the day of its death I would return to the tree I had taken it from. I did that for twelve years uninterrupted afterwards. Every single December I would go back to that tree, which is still there. It became like a tombstone, the tree. The bird was no longer tangible. It had decayed into dust. But I kept going there every year and it was only there that I could constructively contemplate dealing with the loss.

I go to the beech now with Sid and Nancy, my nine-month-old poodles who are giving me the runaround. I remember the first time I saw it. I was with my last dogs – Itchy and Scratchy who I've since lost. We were just over there nosing about and I looked through and there's like an archway you can see through and the tree was standing there and like I said it was just like love at first sight – because it's so huge and so impressive.

Since then I've always gone to it. I went with Itchy and Scratchy, who were a massive part of my life, and we would always go and sit under the tree and we would stop there when we were walking and I'd give them a treat occasionally and we would have a rest – about, from the house, probably about a kilometre. It was a stop for us. When they became ill, we would sort of pilgrimage to the tree and sit and reflect upon the changing aspects of life. Particularly when they got older they would sit on my lap and we would sit beneath the tree and just take ten minutes. Again it was taking stock, really, the three of us, and then the two of us, and then the one of us.

After that, for a while, I only went to the tree at night, in the dark. I didn't go into the woods at all during the daytime. I couldn't face that environment, but I found that I could go into it at night. I would go with no torch, just wandering around. Going there in the daytime felt too close.

The woods were our home. We would sleep in the farmhouse and have our meals in there. But the woods were Itch and Scratch and my home – where we spent all our time.

RUTH BARRIE
Filmmaker

My son Sol was very contemplative. He loved just being out and gathering things. We first noticed there was something wrong with his health when he was four years old and he'd seemed like he'd had viruses all the way through the winter and was quite tired, not really wanting to walk as much. I remember there was this occasion when I was kind of running out the house and he fell on the stairs, and I was in a rush so we just kept going on the scooter. I dropped him off and I went to yoga. It was only when I was lying down, at the end of yoga in a state of complete relaxation. A friend of mine who had died flashed into my mind and I had a strong feeling: "Sol's really ill."

We were really lucky because a diagnosis can take a long amount of time, but it didn't for us. We went to a doctor and she sent us up to Edinburgh's Sick Kids [Hospital] and he ended up having a scan. I remember seeing the way the radiologist stopped the scan. I saw her face, and she said, "We're going to go over and see a doctor." As we walked it felt like the whole world tipped on its axis and my legs would give way. Sol had a brain tumour.

First of all he needed an emergency operation because the reason his symptoms were showing was because the fluid was blocked to his brain and he had hydrocephalus, the build-up of pressure in the brain. At first we thought he had come through it brilliantly. They'd got rid of the cancer and he was recovering really well, but then he had this complication, posterior fossa syndrome, so on the first day he lost all movement and speech.

All in all, it was four years that Sol was ill. The beginning was very traumatic – he had to relearn everything and was in terrible pain with pneumonia and then as soon as possible he had to have radiotherapy and then chemotherapy. But in amongst that was an incredible beauty, seeing the love that Sol's situation sparked in

other people. You're constantly awakened to the preciousness of each moment. We spent a lot of very special time together.

I was able to care for Sol at home when he was ill at the end and William, his dad, and I were with him when he died. It felt a bit like birthing, creating the space for him to relax into death, it was very peaceful. Sol also stayed in the house following his death, for about ten days. It was good to have that time – very powerful and necessary, I think. I couldn't imagine him suddenly being gone.

We had a burial at Binning Wood. It was lovely to do that for Sol. I felt very happy about the way it went. He was in a natural wicker basket. The feeling in the woods was sacred, so beautiful. Sol was lowered into the ground and there was a celebrant who had some special words to say and we all threw in the traditional rosemary and lavender and rose petals. A friend arrived at my shoulder and said that she would like to chant a song. It was a beautiful moment.

She looked amazing. She was all dressed in white. There were rumours that a shaman had been at the ceremony.

Then there was a little moment of silence, and just the wind, rustling through the trees. In that moment, I felt like I could sense the forest move and live and breathe. I felt reassured that I was not leaving him somewhere cold and awful.

The couple who run Binning Wood had asked if we would help with the tucking in, the putting the earth back in, and friends and family took turns. I do love the idea of him going back to the earth. The first time I went back to the woods following the funeral, it felt like I could go there all day, and just be writing, or drawing and writing, being there. You sit there and once the waves of emotion pass through you, you can look up and the beeches are waving above you and birds flit through branches and you feel the forest as a living, breathing entity. I love that Sol is a part of that now.

CALLUM BRAITHWAITE
Tree surgeon

In spite of her insistence she would beat cancer, my mother did make some arrangements in case it went the other way. Shortly after being diagnosed she spoke to a friend who owned a woodland and it was agreed that, if she died, she would be buried there. She had picked out a spot beneath a sprawling cherry tree, a fitting resting place. After becoming an order member of the Community of Interbeing – a Buddhist group following the practice of Thich Nhat Hanh, she was given the Sangha name True Cherry Blossom.

She lived for ten months after her diagnosis, not with cancer, more in spite of it I would say. She was cared for almost entirely at home by friends and family. The community she'd built around her, that she'd always wished for, coming together and supporting her though her illness. They allowed her to live as true to herself and her practice as I believe she ever had. Somehow she radiated life, despite hers gradually slipping away from her.

She died at home in July 2017. As per her wishes, her body was cleaned and prepared for burial in a felt shroud by her closest friends and my sister. Meanwhile we hired a mini digger, dug her grave and wheelbarrowed a couple of tonnes of topsoil to the site, in place of the rockier soil we'd excavated. Three days later her funeral was held in a packed-out village hall. She was taken from there to the woods in a VW transporter, and in the presence of her family and a few friends she was lowered into her grave with an old climbing rope. We said some final words, and then picked up shovels and began the task of filling her grave in. There were smiles and the odd laugh as we worked away. We returned to the hall where there was music, dancing, food and laughter. A friend told me it was more enjoyable than many weddings she'd been to.

Despite the unimaginable pain I felt at her death, I feel like being present throughout made it more of a process; and far from it feeling traumatic, being so intimately connected with this process felt incredibly natural. It made me realise how disconnected with death and grief we as a society have perhaps become. Funeral directors, hearses, grave diggers . . . All these things take us away from those final moments we can be around our loved ones. Burying my mother ourselves brought us together as a family one final time, under the sprawling branches of a cherry tree.

————————————

We don't have to bury a loved one by a tree to find solace in them. The woods are often a place to which we bring our grief, almost accidentally. Trees speak to us so much about the nature of life and death that we can find them triggering these feelings. We see shapes in them, read their stories.

A friend described to me how, after the death of her dad, she went on a guided meditation around Epping Forest. "It culminated," she recalls, "with us being invited to select a spot and talk about it. I found these three trees, one of which was dead, but still standing, and then a younger fully mature tree and then a sapling. It was basically my family. The relationship between the trees became a metaphor, an analogy, for me and my family. When I explained it to the group, I found I was crying."

Many others spoke of how they processed their grief under the canopy. Zakiya here talks of emotions resurfacing around the traumatic loss of both her mother and brother.

ZAKIYA McKENZIE
Writer

When I started going for walks in the Forest of Dean as writer in residence for Forestry England, I didn't expect any of what I felt. I thought it would be the relaxed shinrin-yoku kind of bathing and meditation. I didn't think what it would do would be to churn up things. I thought it would be like a gentle breeze, or a soft wave, but instead it was like the bottom of the ocean, where everything is churned up. I didn't expect that at all – to have any of the kind of introspection that I did.

It was only when I started walking there on my own, once I was under the canopy of the trees where there's no sky, just within this thicket of bush and tree, in darkness, that I was plunged there, into the grief I feel over the loss of my mother.

Since I started walking in the forest, I'm not so much afraid to think the things I used to just push to the back of my head, thinking, I don't want to think about her and what happened to her. Now I tend to write every day – and as if I were speaking to her.

I don't know if grief ever ends, but it definitely has changed for me. I am much more able to deal with it.

MOIRA BARKER
Manager

When we moved into our home as a family in 2004 our three children were at primary school. They wanted to play outside and this was the first time that I couldn't see them directly from the window.

The gathering place for the children was at the bottom of our road and round the corner "under the tree", but I couldn't see the bottom of the tree from the house so I would walk down the road and round the corner to count the children and then return home.

The next summer I realised that the shadows of the tree would project perfectly all along our bedroom window as a living piece of art. When my husband is away on holiday or work I sleep with the blinds open just to watch the tree.

I used to watch the leaves fall in winter and feel sad that the tree was a black shadow and then joy when the leaves returned heralding another season. I have marked time with that tree.

On 26th February this year [2020] our

beautiful eldest daughter Katy slipped peacefully from this world in her sleep, from what was called Sudden Adult Death Syndrome, and I have spent every night watching the tree in my grief. I have watched the leaves return with joy and sorrow.

I watch the tree frame the sunset and the sunrise, but most beautifully the middle-of-the-night skies which I see with wonder as I look at stars and think of the tree connecting us here on earth with the heavenly skies. The connection to the children's childhood and memories is not something that can be taken away from us. Katy sailed through life with only a light touch on material things in life and will forever be in heaven awaiting us. In the meantime I will watch the tree move and change, light up and bow down and celebrate the connection between nature, memories and eternity.

DONNA HASTINGS
Family support worker

One of my earliest memories that involves trees was when I was around eight years old. My dad was trying his hand at photography and we went to a local park that had a loch and was surrounded by beautiful trees. The day is vivid in my mind. The sun was shining and I remember laughing as my dad encouraged me to climb on the tree and the whole time he was snapping away with his camera.

Whenever I look at that photo now, visit the park with that particular tree or see a willow tree which I am particularly drawn too, I am filled with warmth and happiness of a memory from a time long ago.

My dad was diagnosed with stomach cancer when I was fifteen and managed to survive for six years, eventually dying from secondary cancer to his liver when I was twenty-one.

During the six years, times were tough, especially as I was also navigating my way through my teenage years. My dad, David, was a private, proud man and wasn't one for sharing what was going on with his private life. Hoping to protect us, information was limited. I was told he had a tumour but, at fifteen, I didn't understand this meant cancer and certainly didn't know he could die. He opted for no treatment and tried to live the best he could, but it impacted our family greatly.

One of the most difficult things to watch is seeing someone you love, who has always been so strong, protective and proud, slowly deteriorate and become a shadow of the person they used to be. Perhaps that's why I've been drawn to willow trees since my dad was first diagnosed with cancer. There's a symbolism there, of standing strong, proud and protective. That's what I feel when I see a willow. The willow tree is also thought to encourage us to find relief from sadness and grief through tears and expression instead of keeping all our feelings locked away.

LAURA ALCOCK-FERGUSON
Ancient Tree Forum CEO

The treescape that I'm familiar with from where I lived till recently, is the beech woods of the Chiltern Hills. It was a wrench to leave those woods because they were my saviour – through all of those ups and downs of seventeen years of life. I would escape into the woods for runs and walks and bike rides and just thinking time and it's where I found my climbing tree, as I called it. I would go and sit in it, mostly at dusk, when it wasn't my turn to put the kids to bed.

I went there when my mum died recently. It gave me a lot of solace to have somewhere to hide, as well as walking through the beech woods and having the familiarity of the trees. At times in your life when there's something difficult, the familiarity of landscape and trees and that sense of how long they have been there helps. They weren't ancient, they weren't particularly old, but they held a story for me and a consistency and continuity that I appreciated at times of change.

We humans have a habit of planting trees as a connection to those not there. Dame Judi Dench has spoken about how she plants a tree every time a close friend or relative dies – for her brother, her late husband Michael, her friend Robert Hardy.

I married into a family with a similar tradition of planting trees. During a memorable visit to my husband's aunt and uncle, the adventurers John and Marie Christine Ridgway, who live miles from the road end on Scotland's north-westerly tip we were taken on a tour. For there, in this remote, rugged spot, they had more company than it seemed at first glance. Over decades, a hundred and twenty trees had been planted on a steep hillside. Almost like a person disguised in leaf, they were individually named for family members and many others who had helped the Ridgways put down their own roots in this unlikeliest of places.

"Now this is Sibyl D'Albiac, without whom none of you would be here," John Ridgway told our children, as he presented them to a tree representing their great grandmother. "It's is a Great Wellingtonia. You'll see it's extremely thick at the base. In two thousand years, it'll be 12,000 feet high. According to the foresters in California, it will take 2,500 pick-up trucks to take the logs away."

DAME JUDI DENCH
Actor

I have never counted how many trees there are in what I call my extended family of trees that I have planted at my home. They are dedicated to my friends and family, and the wood is ever growing. I am afraid I am due to plant at least thirty more. They all represent people that I dearly loved and were part of my life, so when I look at them the person who I planted them for are very vivid in my mind.

As a young child I was always very distressed to see lorries carrying lots of logs. I was irrationally upset about it. The more I learn about trees the more I wonder about them. One of the greatest discoveries for me, when I made my documentary *My Passion For Trees*, came when I was given a machine a bit like a stethoscope. When I held it up against the trunk it enabled me to hear the rush of water underneath the bark. It was a staggering experience. So now, when I look at trees in the winter, barren and without leaves, it is rather heartening to know what is going on under the bark.

I have said daily how incredibly fortunate we are to have been in lockdown surrounded by trees in the garden and spend time with them when some people have not got such luxury.

HELEN PATIENCE
Photographer

We call this tree "the daddy tree". When my husband Greg died of a brain tumour, not long after the birth of my daughter, I wanted somewhere we could visit together as she grew older, somewhere safe where we could talk about her dad. I thought immediately of this tree in the Botanic Gardens [Edinburgh].

What was wonderful was that when I started speaking to my mother-in-law, she instantly thought of exactly the same tree – she used to sketch it as a young artist and it has always meant a great deal to her too.

The "Daddy tree" brings comfort to my daughter and I. It allows me the time I need to sit in nature and it gives space to the feelings I have over our loss. I've shed tears here but also smiled remembering visiting the tree with my husband. My daughter plays happily under the branches, often singing while collecting pinecones. It's a safe space where we chat about Daddy, or any difficult feelings, protected by the branches above.

KIRSTY WARK
Television presenter

My parents adored gardens, so I spent a lot of time in the garden. There's a language of flowers and trees, heartsease, rosemary for remembrance. These are all culturally important. I also think because of the themes of growth and renewal, plants can take a place, for me anyway, of religion. I'm not religious, but actually you can express a spirituality in what you're interested in planting and the trees that you like.

In my last book, *The House by the Loch*, I featured the mountain ash, because it's a species which is planted in burial grounds as it was meant to protect the spirits of the dead. My character is desperately trying to make sure she gets the exact tree she wants, the Joseph Rock Sorbus, and has a row with the horticulturalist at Threave Castle. That scene also reflected the way what you do when you're bereaved is to have whatever agency you can. I remember this from my own parents' deaths, that what you do is you control things that you can.

We planted a rowan on the loch where my father loved to fish, in remembrance, down in Galloway near Carsphairn. That seemed right for him and also because old houses in Scotland most often have a rowan at the gate to ward off evil.

I do go back to my dad's tree. I sometimes, especially in spring and summer, take a picnic down to the loch. You go back at different points in your grief and it's so reassuring to find that things don't change. So if I go down to the loch and I sit at the slipway where the boat goes in beside the tree and I hear the peewits, I think that well, nothing else has changed since I used to go out on the loch with dad.

When Mum died, we planted trees for her. I think trees are incredibly comforting things, especially tall, big trees. There's something so elemental about them in the landscape. To me they are a living memory. They are also a way-marker. Sometimes a particular tree will tell you that it is time to turn off a road, or tell you that you are at your destination.

Vicky Allan

THE HEALING TREE

But trees don't just help us connect with lost or distant individuals or families. Memorial trees even have a place in the aftermath of our darkest hours. In Manchester in 2018, the year after the Manchester Arena bombing, the city created a trail of Japanese maple trees from which people could hang their messages of hope, sympathy and bereavement.

At Ground Zero, following the 9/11 attack, astonishingly, a pear tree was discovered in the rubble, roots snapped, branches burned. It was rehabilitated and replanted as "a living reminder of resilience, survival and rebirth".

And after the Grenfell Tower fire, too, the local community planted trees around Al-Manaar, the Muslim Cultural Heritage Centre. Rabbi Natan Levy took part in the interfaith project and recalls:

"These trees are one component of a larger matrix. They are one of the catalysts, a bit of the glue that brings people together, around tree-planting. And Al-Manaar was itself a place of real healing. A lot of what happened there wasn't just about the fire, but the feeling that the community had been left behind and didn't have a voice."

Resilience, hope, survival, rebirth . . . these are the themes that whisper through the leaves of trees. We come to them for comfort – to be reminded of how we can still endure, and also, how, in the end, we are all an elemental part of something bigger, to which we will go back.

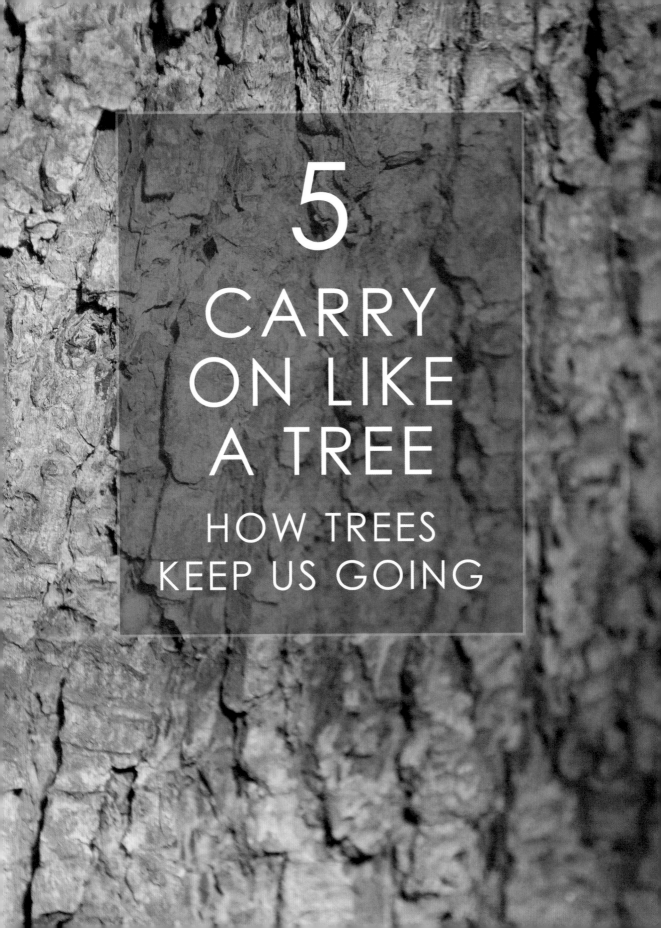

5

CARRY ON LIKE A TREE

HOW TREES KEEP US GOING

I am still alive — in fact, in bud.

KATHLEEN JAMIE, 'The Wishing Tree'

Trees are a comfort. They can sustain us, keep us going, give us hope. And they can gently distract us.

We have known for decades that a view of a tree from a window can enhance our healing from surgery. In a much-cited 1984 study of people recovering from gall bladder surgery in hospital, by the psychologist Roger Ulrich, it was found that those who looked out on leafy trees recovered faster and needed less pain medication than those who looked out on a brick wall. Interestingly, Roger had first-hand experience of this. As a teenager, he'd had serious bouts of kidney disease and spent long periods at home in bed, looking out the window at a big pine. "I think," he has said, "seeing that tree helped my emotional state."

The body of research around time spent in leafy places is growing. As psychiatrist Charlotte Marriott describes, "Green space exposure has long been shown to be beneficial for health and wellbeing, reducing the risk of type II diabetes, cardiovascular disease, high blood pressure, stress, anxiety and depression and improving sleep. Some of the mechanisms include reduced levels of the stress hormones cortisol, adrenaline and noradrenaline as well as the impact of phytoncides, volatile organic compounds released by trees, on the immune system."

But, when we focus on the research, what's often not discussed is the texture and subtleties of what trees mean to us. During the 2020 pandemic, people across the UK have reconnected with some of that meaning. They have shared images of the trees they can see from their window, in Rob McBride's project #treesfromwindows, or competed on twitter for former Labour aide Alastair Campbell's "tree of the day" accolade.

During the same weeks, as I watched these snatches of solace and hope shared on social media, I was also listening to people tell me stories of how trees and greenery had featured in their recovery from illness, or sustained their spirits through it.

IONA MALCOLM
Garden shop owner

I had an accident in December 2019. I was crushed by a van and trailer and it left me with a haematoma in my leg, and I had to rest up. When that happened, I thought, Well that's hard, but I just have to be patient. And then, not long after, I got a cancer diagnosis and I felt, Oh dear God, what now? I felt like Rasputin who was shot and stabbed and poisoned and drowned.

I found the lump when I did a self-examination. My GP couldn't feel anything but referred me to Raigmore Hospital, where the consultant also couldn't feel anything, but sent me for an ultrasound and mammogram.

Iona Malcolm

That showed that there was something there. When they did a biopsy, sadly it proved to be cancerous.

The past few months I haven't been able to get out because of the lockdown and that's been really hard, because even before this, I was confined to the sofa with my leg elevated because it was injured so badly.

Seeing the trees outside has helped. One side of our house looks on to the High Street, but I have a lovely bamboo outside the window, and the room I am sitting in now looks out on to the garden and the churchyard where I can see the evergreen trees; Scots pine and cypress, and then the deciduous ones closer to me. I just love looking out at it. My favourite combination of colours is pine trees against the blue sky.

I remember being thirteen and we'd gone on our first foreign holiday to France and we were cycling through the pine forest near Bordeaux and the sky was a dazzling blue and then the greenery of the Scots pines, just lifts your heart.

Being outside is something I really need. When I was twenty-five years old I first started showing symptoms that would later be diagnosed as multiple sclerosis. It was my eyesight that was the first sign. It felt like there were pixels missing and I was referred to the Eye Pavilion at Edinburgh Royal Infirmary. They said I had optic neuritis and gave me steroids and sent me home. But there followed an accumulation of other neurological issues, and I was referred to a neurologist and sent for a scan, which proved inconclusive.

This all happened in 1999; episodes followed which became more and more severe. We moved to Kingussie in 2005 and I continued to exhibit minor symptoms, but in 2009 I had a major MS relapse which meant I lost the ability to walk. That was a tough spell, but if there is a bright side to it, it was that I got a conclusive diagnosis of relapsing, remitting MS and so could begin treatment.

After the diagnosis, I went to work for the

Laggan Forest Trust, a small community charity which bought a section of forest just outside the village of Laggan.

I was lucky enough to be outdoors a good deal of the time and I was probably getting all the vitamin D I needed from the sunshine. There's a track there that takes you on a twenty-minute circular walk. Walking that, after my relapse, was a really big thing for me because I was learning to walk again. I was trying to build up my strength to walk greater distances and it was a huge positive step for me when I managed to do a full circuit.

One of the things that has helped me is not only trees, but gardening. I was pretty depressed when I first got my diagnosis. A turning point came when my husband cleaned out the greenhouse and challenged me to make something of that. I might not have been growing something as big as trees, but I started sowing vegetables and it was just so lovely to see the tiny leaves come through the soil. I became so obsessed I ended up setting up a gardening shop.

LORRAINE McGUIRE
Child development officer

After my mastectomy in January 2020, I couldn't travel very far but was able to get out to Kelvingrove Park, our local park. I would sit on a bench underneath a big tree on the hill and watch my daughter feed the birds. After a few outings we were able to distinguish between the crows – we gave them names; Mallen, Grubber and Tina were the three that came almost daily. The magpies stayed clear until the crows had their fill.

I love the park because it gives me a touch of nature when I'm unable to get into the wider natural world. I'm sort of Pagan, so nature has always been important to me. I feel like I can breathe fully when my back is against a tree and I'm looking out towards a landscape where the mark of mankind is minimal.

Since starting chemotherapy, I am now in the "at risk" category so I'm only allowed out to go to the doctor's or hospital for treatment. This means that my connection to the natural world is tenuous and restricted to my balcony. I go out there first thing in the morning and listen to the birds. There is a tree opposite my house where a pair of crows have taken up residence. The female has been nesting, I think the egg(s) have hatched – although I haven't seen the baby. The male crow visits my daughter when she goes outside, because she has been bribing him with monkey nuts.

Today I went to the doctor's and the male crow followed me along the road, cawing. It sounds a bit mad, but I think he recognises me from the balcony – and even if it is mad, I got a little peace knowing that I can still interact with nature, even if I am housebound.

ISABEL McLEISH
Artist and forest therapy guide

In my final year of my Contemporary Art Practice degree, I became extremely unwell and was hospitalised with acute anaemia.

During my recovery, as I struggled with fatigue, I began to explore art practices which incorporated mindfulness and had a meditative aspect to them as I hoped to find a means of creating work which wasn't exhausting.

My mum also took me on a daytrip to Glen Affric in the Scottish Highlands so that I could appreciate the changing colours of autumn and get some fresh air with minimal effort! Taking photos from the car and simply admiring such beautiful native highland woodlands really stayed with me.

The physical and mental toll that the experience of anaemia had on my overall health was a challenge. I needed to work at a much slower pace than before and found that a sense of immersion whether in the natural world or a specific activity was actually energising.

A symptom of acute anaemia is breathlessness and I began exploring the notion of "breath" in my art and considered the links that this has with trees as they oxygenate the air we breathe.

THE GARDEN OF RECOVERY

The pandemic period of 2020 has been one in which many of us have turned to trees.

But, for the explorer Robin Hanbury-Tenison – who, after contracting Covid-19, spent weeks on a ventilator – an oak tree through the window, and a visit to a hospital garden, were about more than mental soothing. They are what made him feel he would live. Robin was lucky enough to have been at Derriford hospital in Dorset, whose ICU department was pioneering taking such patients out into a garden.

Kate Tantam, a specialist in intensive care, is the driving force behind this innovative project. In a newspaper interview, she described how intensive care needs to be humanised. "The garden is part of the humanisation process. We take patients outside, even if they are ventilated, so they can feel fresh air. There's a power in the sunlight and fresh air and rain. Some patients like to be out in the drizzle."

ROBIN HANBURY-TENISON
Explorer

I caught Covid-19 early in the epidemic and at the age of eighty-three. I ended up in Derriford hospital for seven weeks and was in an induced coma for five of them. The worst thing about it all is that being immobile for so long is so boring. That's my strongest memory – the sheer boredom. I quail at the prospect of going back into hospital, even though they were so wonderful there.

Luckily I was in a wonderful ward on the top floor, where I had a huge window looking out on to a big oak tree. I was so relieved to be looking out at a tree rather than a blank wall. Things happened out there. Buzzards flew past. Birds flew in and out of the trees. Nature is deeply interesting and one can lose oneself in that and feel better. This was the month of April – and the tree was in leaf.

I never really thought I wasn't going to make it. There was a consultant who came at the beginning, when I was still compos mentis, before I was put under. He arrived with his entourage of nurses all round me, and face masks on. He said, "Hmmm . . . doesn't look very good. You've got two choices. You can stay here in this nice ward looking out on the oak tree, very comfortable, in which case you will almost certainly die. Or you can come down to ICU. You can have horrible things done to you, tubes stuck into you everywhere, and you will then have a 20 per cent chance of living. You can choose."

I said, "Don't be ridiculous, of course I'll choose the 20 per cent chance, over certain death."

Around about this time I wrote to my family, saying, "Bugger. It doesn't look too good, does it? But I'll get through." I sort of believed I would.

Key to my recovery was this glorious occasion when I was wheeled out into the sunshine and into a tiny little healing garden that they have at Derriford. That was the seminal moment in which I actually decided I was going to live. They had only just made the garden and I think I was only the second person to go into it. But when I went out, there was something about the very fact of not being in an intensely technical ward, with bells and whistles going off all round, tubes everywhere and bright lights all night,

suddenly being in the fresh open air and with the sunshine hitting my face.

Before I was wheeled out into the garden I'd been through some pretty unpleasant, hallucinogenic times.

I saw snakes crawling over the bottom of my bed, big cats that I could reach out and stroke, bats climbing up my saline drip. I was very aware of circles of other existences. There were some quite zany worlds, including one where I had the ability to switch off the world, because the world was in such a bad way. But I didn't because that would have also switched me off. I remember thinking, Oh well, give it another chance.

It's been a month and a half since I came out of sedation and I'm already 80 per cent fit. I can walk around five hundred yards now. Where I am now, on my veranda looking out at the wonderful view we have from here at my farm, Cabilla, mostly of woodland, is very therapeutic and I spend a lot of time looking at it. I have three trees in particular that I watch throughout the seasons. They are elms. The elm is a particularly nice tree, which varies through the seasons. They are my pride and joy. I dread the idea that an elm disease beetle will fly in and kill them. They really are friends.

HEALING GIFTS

It's not only the view of a tree, or the breathing of its gases, which can help us heal. The trees bring us other medicines. A key component of aspirin is salicylic acid, originally derived from willow bark. Tree medicine is a branch of herbal medicine. Here Emma Roe describes the way she hunts for her treatments.

EMMA ROE
Medical herbalist

I watch the trees every day. I find spring particularly exciting; I watch for the new leaves to start poking through – the birch, the hawthorn, the oak – these leaves, tender and soft, are less bitter and astringent in the spring, perfect for medicine making. Next come the flowers: the hawthorn – picked just before opening, I pick enough to dry and to make into a tincture (alcohol-based medicine); the elder – just in time for hay-fever season; and the lime flower – perfect for a relaxing infusion on a summer's evening. Then the fruits – elderberries, horse chestnut, cherry, apple.

My family is kept busy from spring through to autumn. Luckily, we have winter to rest. Making medicine is a family affair – my husband is also a herbalist and our four-year-old is learning.

MEMORIES BLOSSOM

Early on in our talks about trees, Anna would often mention her much-loved grandmother who passed away in 2019. She had, in her final years, been suffering from dementia, and Anna, who would take her out for walks, had noticed how, often, when she was outside among woods and trees, she would blossom again.

This chimed with other stories we were hearing about the ways woods might stimulate and bring joy to those with dementia. That qualitative impact is something that researcher Mandy Cook has studied. Working with Scottish Forestry's Branching Out programme, she looked at the impact of time spent in the woods on those with early-onset dementia, and found many carers and participants testified to a positive effect.

Being in the woods even stimulates memory, as one dementia care consultant working on the project observes. "You can feel that chill, feel that fine drizzle, feel that warm breeze and you can feel the sun. That stimulates reminiscence and it can create conversation and that's engagement and interaction. Woodlands have a huge resource, it's like a library of stimulation."

ANNA DEACON
Photographer

This time last year, when the cherry blossoms were full to bursting, scattering thick layers of pink petals over the paths and flying like confetti on the breeze, I took my granny for a walk. Her nursing home looked over the park, but along with her dementia, her eyesight was diminishing so gazing out of the window wasn't something she could enjoy any more. But to feel the tangy, salty sea wind on her face, as she was wheeled along the avenue of blousy pink, to be able to sit under the bough and try to catch the petals as they rained down on her was a sensory experience that took her out of herself.

In the same way that, right up to the very end, Granny would be moved to conduct and sing along to classical music, knowing every single note, beat and melody, being outdoors seemed to awaken something deep inside her. Despite her memories, and words, falling away, these things were inside her and right up to the very end they kept her being her. The essence of Granny was still there and this time outdoors helped unlock it.

Those last months with Granny, were so precious. Our roles reversed. I remembered things she gently used to show me in her own garden and I tried to do the same with her. Once the futility of trying to reminisce, or talk to her about anything apart from what was right in front of us in that moment had passed, and acceptance that this was the new normal sank in, we settled into a different kind of relationship.

We would wander around under the trees, picking up feathers, pine cones, flowers, leaves. Often, we would sit and just look up at the trees, searching for but not finding words, instead holding hands and watching the birds flit about, hearing the leaves rustle, distant shouts of children playing and dogs barking. Sometimes we would go and touch the trunks of the trees; Granny would run her long pianist's fingers over the knobbly bark in childlike wonder.

TIM CHAMBERLAIN
Director of Wild Tree Adventures

Mum died at the end of February 2020, just before everything went into lockdown with the coronavirus pandemic. We were very sad and yet also somehow so relieved she went before everything happened with the virus, particularly what happened with visiting rules in care homes.

Mum and Dad used to live in a lovely little village and when I was a kid my dad helped organise the village to plant trees. Many years later, on one of our many walks, Mum told me that she had a willow wand left over from planting on a nearby nature reserve and she stuck it in the ground. And it grew. We used to walk past it nearly every day as we walked the countryside, finding pleasure in small natural treasures. Later, when her illness stopped her talking, these walks became the real magic of living with dementia.

I was away from home when she was diagnosed with dementia back in 2002. Dad looked after her for a while until he died suddenly in 2005. My initial plans to travel the world after art school were put on hold to go and look after Mum.

Mum stopped driving soon after that, so we had a while where, because I didn't drive, we just cycled everywhere. There was lots of activity, fresh air and walks, so many. We went hunting for mushrooms in the woods and later when I could drive, we walked the beaches and cliffs along the coast. Hard as it was, I remind myself that it was the illness that gave us that magic time together, and all those adventures. We even made it out to the wilds of Labrador!

For a mum she was pretty cool. We visited so many wild corners of the Scottish Borders and hiked in sunshine and deep snow. There was even one winter when we went sledging twice a day for two weeks: she had Alzheimer's, and she was living with it.

The willow was a really lovely marker in the landscape. Before she went into full-time care, we used to go past that willow and say hello to it. Once winter I took pictures of her in the snow in front of it. It's lovely for me, now that she has died, to know that tree is still growing. Last time I went back, I had a climb in it and thought of her smiling, by her tree.

Giorgos Tsiris

THE LAST LEAVES

Trees, people have told us, are a comfort at end of life, too. They bring meaning and solace. For instance, at St Columba's Hospice in Edinburgh, where Giorgos Tsiris, as arts lead, helps patients in their final months. He believes the beauty of the gardens and trees there offer "opportunities".

He says, "Trees can put life in perspective. For people at the end of their life and for their families, trees can offer an image for reflection. They can offer an opportunity to pause and to think. To think creatively, imaginatively and perhaps existentially."

One of their biggest comforts, he notes, can be the way they offer us metaphors for "our own lives, our own connections with the world, and even what might be beyond the material world". Tsiris sees that at work at the hospice. "It is not only the visible parts of a tree – the trunk, the branches, the leaves. It is also its invisible roots that ground the tree to the earth, that give stability, connection and strength."

6

THIS TREE SHALL NOT BE MOVED

THOSE THAT FIGHT

Who robbed the woods,
The Trusting woods?

EMILY DICKINSON, 'Who robbed the woods'

"Tree sitting is a last resort. When you see someone in a tree trying to protect it," wrote Julia Butterfly Hill, one of the world's most famous tree sitters, "you know that every level of our society have failed, the consumers have failed, the companies have failed, the government has failed."

The fight to save trees has a long history. Newbury. Pollok Free State. Twyford Down. Sheffield. Here in the UK, the names of some of the places where people have made a stand, living in trees, refusing to leave, sometimes chaining themselves to them, have resonance. They are, in a way, historic battlegrounds. The first ever tree-sitting action took place in 1978 in New Zealand and led to the Pureora Forest Park.

Tree sitting is still, even now, what people often turn to when they want to protect a tree or trees. It's what, over the last year, people have been doing at woodland sites across the UK where the HS2 high speed rail network is being built.

The fight for trees is often waged on two fronts – physical protest and the law. When I spoke to Chris Packham in spring 2020, he was in the midst of a legal battle against HS2. An injunction he had filed to pause the works had failed, and, even in the midst of the Covid-19 crisis, contractors had moved in on some of the sites. Chris was pursuing a review of the project on the basis that the Oakervee Review failed to quantify and address the full impact of HS2's likely carbon emissions. He was upset not just for the trees, but for the wildlife that inhabited the woods.

"This is, of course, the bird breeding season and the bat roosting season – which has not yet come to an end," Chris said. "They're in the process of moving their roosts from one to another."

The people who fight to save trees often spent weeks and months, even years, in them, and developed an intimacy with the trees that few of us can imagine. They sleep in them. They dream about them. They breathe their air. They become part of their wildlife. Some tell their stories here.

INDRA DONFRANCESCO
HS2 protestor and tree protector

I didn't actually come here to be part of evictions. I came here to train some activists and got caught up in the Covid-19 no-travel situation. Then I realised actually being here was really important. I feel like it was fate for me – right place, right time, fortuitous or terrible, whichever way you look at.

These works are set to destroy or damage one hundred and eight ancient woodlands. It is very much a magical fairy kingdom here, with woodlands that are complete and intact and functional. This used to be the Arden Forest of Shakespearean fame; *Midsummer Night's Dream*, that kind of thing. What's left is pretty much just wildlife corridors, but they are still wildlife corridors. There are wood anemones, carpets of them, indicators of ancient woodland that you just don't get anywhere else. Also, the soil is incredible. It's perfectly broken down, moist particles.

I was part of the original road protests, from Twyford Down onwards, Salisbury Hill, Newbury – which was tremendous. It was a critical mass of people, and the exceptional thing is that there were people living in trees, being in the woodlands. It has this effect of re-wilding you.

What's happening here is so wrong. It is wrong on so many levels. These trees do not deserve to go. These trees are so important. They are not only a wildlife corridor, they are an existence. They are here and they are existing and to see these ignorant men with all their machines out here – this is corporate capitalism and patriarchy and everything boiled into one.

The protest at Newbury was something else. That turned me. Even though I'd been involved in tree camps before, that was the biggest. And I remember we were being chased by security and we just stopped this digger from coming in and it was just pulling everything up. I managed to throw myself under a tree that was half exposed, with its roots, and I held on to the roots.

I was under this tree and I just felt the most overwhelming feeling of calm and peace. What I realised I was feeling – and then I consciously tried to do it – is the trees' energy fields. Trees know when other trees along the line are being killed or whatever. I felt those trees were sacrificing themselves and it was such an intense feeling of being near something so old and so ancient that was getting ready to die.

And you can feel it. It's like they show themselves. I think that it was a big turning point for me, Newbury. Because, you know, we were there doing what we were doing because that was the right thing to do. We knew about climate change, we knew about ecosystems. Factually, theoretically and academically.

But actually to feel it was something else. I suppose it's like being in the jungle and you're sitting there and suddenly a gorilla comes up to you and touches you and you just freeze.

TALIA WOODIN

Photographer

Both my parents were Green Party politicians and I grew up with stories about Greenham Common and Newbury bypass. When I was seven, I was saying I want to be a Greenpeace photographer when I grow up. I am now working as a photographer for XR [Extinction Rebellion] and living my childhood dreams.

Trees were a very big part of my childhood and I grew up with stories of my parents protecting trees. My dad died of lung cancer when I was four. He had never smoked and was only in his thirties. It was hugely shocking. But we lived in Oxford which has one of the highest concentrations of air pollution in the country. An average of 64,000 people die every year from air pollution just in the UK. I believe my dad was one of these people.

Newbury was burnt into my mind as this iconic time – I'd heard all these stories and I've always been itching to protect a tree. Then HS2 happened and it just kind of fell into place really that I would do this.

There are five or six camps. Some of them are on private land and are more secure, but the one I'm at now – as Denham is on injuncted land which we're going to be evicted from any day now – so it's a bit more chaotic.

I think it is really important that something like this is photographed. HS2, at this moment in time, is not getting any media attention because there is so much going on in the world. I feel outrage and hurt witnessing the destruction and what is happening and knowing that so few people know it is going on.

One of my favourite images is of an activist holding a sapling at Crackley Woods protection camp up near Coventry. We went on to part of HS2's injuncted land, in which they were clearing the way to build a road in order to transport machinery through. We'd seen that there were saplings along the path that they were completely destroying and a group of us went out and dug up the saplings so we could replant them somewhere safe.

GEHAN MACLEOD
GalGael Trust founder

There was, in the Pollok Free State camp that we set up to stop the M77 being built through public woodlands, a 200-year-old beech, down near the edge of the wood, looking out across the farmer's field. It's still there now. There are lots of memories associated with that tree for me. It was somewhere you could go for a little bit of quiet time. We would have a fire down there quite often. I remember one time my late husband, Colin Macleod, was needing a bit of time away from all the people, so he took a book and a hammock and climbed right to the top of this tree and strung his hammock out and sat up there reading a book for a couple of hours. There was quite often a rope swing hanging off it. When our daughter, Tawny, was born, we hung a baby swing there.

It was also a special tree for Colin, before I met him. I know that some of the ideas that were formational in both the Pollok Free State and GalGael came to him when he was sitting

at that tree, staring out at the field, reflecting. He tells the story of how he had a pile of ecology mags and was sitting at the base of the tree figuring out what he was going to do in response to environmental issues. Then a pigeon shat all over his magazines, which prompted him to think that this wasn't the way we were going to reach the communities round about and pushed him towards carving as a means of communication, really playing on people's curiosity.

The big thing that shifts is that when you're living among trees, and when you're living outside in general, you start to tune into the daylight. You get a sense of what time it is, just by the quality of the light.

I was there about a year and a half, and Colin was there longer. You're lying in the woods at night, listening to all of these branches creaking and there is part of you that is tuned to whether something might fall – because there was a few times down there when a major branch would come off a tree. You were always aware of the noises, the squirrels and the treecreeper birds the tawny owls at night. You were aware not of trees in isolation, but as an embedded part of the environment that you're living in.

I feel humanity is now losing that sense of connection with the natural world, but also that connectedness that trees seem to represent, through their branches reaching out and their roots. Trees seem to inhabit a bigness and connectedness. You feel, here is a being that's not operating at the same pace timewise as you are and whose connections are many.

SIMON BRAMWELL
Extinction Rebellion co-founder

When I was very young, nature was a friend to me at a time when I didn't have many friends and needed solace, especially when my folks split up. I used to spend a lot of time up in trees, either reading or just feeling sorry myself, or thinking or playing. I was very fortunate to grow up when I did, in the fact that I could quite often run off, escape, go and build dens, climb trees and just be in nature as a playground.

Around about 2002 the realisation of my own part in nature really came home to me after I nearly lost my left arm in a farm

accident. I was working on a hop farm. This piece of machinery, a hop press, caught me and dragged my arm through a very small gap. I had to have twenty-three procedures over a couple of years, got dependent on meds and had cognitive behavioural therapy, but it was spending time in nature that brought me back to myself and out of that internalised pit of despair.

I started spending a lot more conscious time in nature. I adopted a practice called a "sit spot" where you would just sit in nature and observe it and reduce your human baseline of noise. That started opening my mind a lot more. Badgers start coming up to you, foxes come across your path.

I've been involved in various strands of activism, and in 2015, I took part in a protest at Stapleton Road in Bristol. We were there for six to eight weeks in the depths of winter, trying to stop the trees being chopped down to build a bus lane. Spending time in an oak, you see what an entire nation the tree is – all the things that live on it. I slept in it, although I hate heights. The trees were quite close together and sometimes there were twenty people up there. We'd have common experiences, similar

dreams. On a few nights we would come down in the morning and everyone reported deep fits of crying during the night. You start to realise how deeply in communication trees are with each other and the many ways they display forms of intelligence.

Eventually we got evicted from the trees. There was a massive sense of loss, frustration, impotence and rage, and that just got me thinking about how activism works, and whether we could have engaged more with the local community. What happened at Stapleton Road in 2015 was this deeper realisation that the kind of activism we were doing wasn't working.

Shortly after the eviction I moved to Stroud to look after my mum who was becoming increasingly frail, which was when I started hanging around with Extinction Rebellion co-founder Gail Bradbrook. We became romantically entangled. We started talking more and more about what needed to be done. Why wasn't activism working? Why weren't more people getting involved? Then we found out about Roger Hallam and went down to see him at a workshop in London, and that was where Extinction Rebellion began.

LUCY POWER
Trapeze artist

We went down, as a family, to London with Extinction Rebellion dressed as forest circus characters and we were on stilts on Waterloo Bridge. Bella, my daughter, was hula-hooping and her picture was everywhere across the papers.

JIM HINDLE
Newbury protestor

The first time I saw Middle Oak, a tree I would live in and fight to save at Newbury, we clocked it as we were walking to our camp, and I didn't actually realise it was on the bypass route. It was quite an iconic thing spreading out in the middle of the field.

I had lived in a tree the previous year, nearby at Snelsmore. It was a very strong experience, very vivid. There's at least thirty feet between you and the earth. In common with a lot of people I had very vivid dreams when I was in the trees. I think we all fiercely anthropomorphised the landscape during the campaigns, but you couldn't deny there was a very strong feeling that you were in the arms of another being.

Newbury became the focal point for campaigns that had been going on for years. I think early on I was of a mindset that the decision had been made. I thought this place is going and all we can do is slow it down and can make the point. They built the bypass at Newbury, but the wider programme was called off to a massive degree and I think it did push that level of road-building back for the best part of a decade if not longer. It never went away, but roads that were on the cards were called off at that point.

Middle Oak was unoccupied. No one wanted to live there. We'd put a platform up, but it was pretty windswept and isolated. Then, the day before works started there was a tip-off that they would use that field by the A4 for a compound, so we effectively needed someone to go up and keep a look out. And I did it because no one else wanted to do it. I put a tarp on the

hazel frame we put up and I was there for the next three months.

In the run-up to the eviction at Middle Oak, I was anticipating being up there. I'd been through an eviction at Stanworth and that was very fraught. I was concerned about my psychological health.

The summer after Stanworth I'd been hospitalised and I felt that going through another eviction would have a profound effect. So I was not looking forward to that, but I didn't feel I had any choice about it.

Middle Oak is still there – it's in the middle of an interchange. I remember the day we found out it could be saved. One of the locals, during the height of the evictions, had been looking at the plans and worked out that it was going to be in that interchange. It was just very fortunate that some people came on board who were able to go down and talk to the under sheriff, who was working on evictions nearby.

I did feel grief over the other trees that were destroyed at Newbury. For me it manifested in my mental health; for others it was a lot of alcoholism, drug abuse, other people with PTSD, so it was quite harsh and arguably a difficult style of protest to sustain.

I think the thing that stays with me most potently is having borne witness to those places being destroyed having got to know them so intimately. Those landscapes were so unique. We witnessed them going.

But it's not only these big sites where the battle for trees is played out. It's also in a more mundane way there in the people who ask for tree protection orders, who sign petitions, who write letters objecting to tree removal. All these things count.

KAREN AND CALUM BENNETT
Graphic designer and structural engineer

We are trying to save a tree right now – a semi-mature sycamore that is one of the last remaining tree protection order-listed trees on the Wardie embankment in Trinity, Edinburgh.

It would be devastating to us if it was removed. On a personal level, it frames our rear garden looking out over the River Forth below. From a wildlife perspective, we see every type of creature enjoying its wonderful position and setting. Collared doves, wood pigeons, chaffinches, goldfinches, robins and magpies frequent daily.

The fight to save this tree arrived in 2019 when a development company purchased the site next door and applied for permission to remove the tree. After failing to secure permission to fell the tree via planning (because of an ownership issue), a planning application was submitted with a basement for a residential dwelling proposed within two metres of the sycamore. Undoubtedly this basement would kill the tree's root system, requiring it to be felled.

During our fight to save this tree, we have fallen more and more in love with it. Researching its history, considering the species that inhabit it, and meeting like-minded members of the community who feel similarly, have all been major parts of our journey. We believe this particular sycamore, and many other parts of Edinburgh's historical plantations, are what help define such a beautiful city.

YOU CAN'T BRING BACK A TREE

Often, tree sitting doesn't work. The protestors are extracted, the trees felled, the ecosystem lost. Too often, in other words, protestors are witnesses to the loss of the thing they tried to save. At Sheffield, 5,500 trees were chopped down and replaced with saplings, before the city's council backtracked and acknowledged it was possible to keep the trees. The protest is considered a success, but many trees were lost.

The fact that it is so easy to fell a tree, or even a whole woodland, and regret it afterwards, is one of the reasons why there is currently much discussion around whether trees, forests or other ecological systems, should have inherent rights. That debate is happening both here and in other countries, too. When people look for tools for the protection of organisms and ecosystems we quickly come upon the idea of rights, of personhood, and what it might mean

if we legally applied them to trees.

Those ideas were there when, two years ago, the citizens of Toledo created what they called the Lake Eerie Bill of Rights. The lake, which had suffered from deadly algal blooms caused by run-off pollution, they said, should have the right to "to exist, flourish, and naturally evolve". Their bill of rights is a document that forms part of a gathering movement, a push for more protection for not just individual organisms but for wider ecosystems.

Are we likely to see such rights in the UK? The lawyer Paul Powlesland offers his thoughts. "I don't think it's likely that trees will get rights any time soon, in the way of, say, parliament passing a tree rights act, because it's so inimical to how our system works and what it values at the moment. But the law is just an expression of politics and power and it's a kind of truism that if enough people rise up and wish to do the world differently, eventually that will bring about change."

PAUL POWLESLAND
Lawyer

In February 2017, I decided to post on the Facebook group "Britain's Ancient and Sacred Trees" asking if there was a group for lawyers to give their time for free to help save trees, assuming that there would be, but there wasn't. However, somebody suggested that perhaps the people in Sheffield might need my help.

Soon I was talking on a phone with someone from the protest who was saying that they were all being arrested under trade union legislation and that the campaign was likely to collapse because people just kept getting arrested. "We think," he said, "the section they're arresting under is not the right one. It's unlawful. We need advice from a QC who is an expert in the area by Monday."

I told them I wasn't a QC and I'd never looked at this area before, and that I was just about to go on a walking holiday in Scotland, but I would give it a go. As I researched, it quickly became apparent that the police had completely misread the section on interfering with a workman's tools. Crucially, the interference had to be unlawful and simply standing in your own front garden or on a pavement is not unlawful.

I remember sending my advice at midnight on Sunday, and going out walking on Monday. When I got back, I saw the reaction it had had. Law never normally happens like this – especially when you're trying to protect nature, which has very little protection in this country. But in this instance they handed it over to the police and they went back into their car and they read it and then came back and said, "Right, we're going to go away and seek legal advice about this." The police didn't come back. Presumably because their lawyers read it and they thought, Oh shit, we missed out the word unlawful.

Two years later, the people who were arrested got compensation for their arrest. What happened in Sheffield is that now it's been shown that the council didn't have to chop those trees down. It was completely unnecessary. But there's no point being proven right years down the line, because the trees are gone. And this is something that the current legal system doesn't really bear in mind – that you need to question first, before you destroy nature because a lot of it is irreplaceable.

It's the proudest thing I've done. That advice combined with the efforts of the protestors to save lots of trees. The advice itself was not the most important thing; it was the brave people standing underneath the trees that made the real difference. But equally, were it not for that advice, then many more trees would have been lost and there's a specific tree up in Sheffield called the Owl Tree, which was the next tree in line to be chopped. Thousands of trees have been saved, including the Chelsea Elm, the Vernon Oak, the Owl Tree, the Abbeydale Park Rise cherry trees.

I'm always keen to emphasise that lawyers who wish to protect nature on a grassroots level should work in tandem with local protectors and activists. Because, at the moment, the law by itself can often do very little. It can be a shield not a sword. We need lawyers to protect protestors as much as possible from the law, but actually the law itself is often not going to save the trees. In Sheffield, the council legally had the right to take down the trees.

That one bit of legal advice was successful, but other than that I lost every single case I did for the Sheffield protesters. But there's tactical learning there in what it is to lose and what exactly is lost, because even though I lost every case in a legal sense the tree campaign was completely successful. It was one of the most successful grassroots environmental campaigns of the last decade. Every time they took someone to court the campaign got stronger. More people would say, "It's outrageous, what are you doing, I'll stand in a tree as well then." The protesters lost every battle, but they won the war.

ANNA GREAR
Professor of law and theory

If you want to ask me about trees, I think they are incredibly ancient, highly evolved forms of intelligence that have enormously complex, networked relationships in which you can see patterns of care, awareness, self-defence, competition. And that's not the way the legal system looks at trees. Very often people will, when they think about tree rights, think about the rights of an individual tree, but I prefer to think about the rights of trees as living, connected systems as well. There's something collectively important about what trees do and are in relation to everything else. If you were to say what's more important to planetary life, trees or humans, you'd have to say trees.

Should we have rights for trees? We have to understand what rights are and what they do; rights are very complicated ideas. We've lots of theories about what they are, but the simplest way to think about it is to see rights as a kind of language and tool for expressing vocabularies of ethical concern and claims against injustice.

That sort of energy around rights perhaps first emerged in the French Revolution. Over the years there has been a long history of groups going, "Hang on a minute, I'm not included in this." We've seen this absolute explosion of various rights aiming at specific, excluded groups. The rights of women, the rights of the LGBTQ community, the rights of nomadic peoples, the rights of indigenous peoples.

Where does this fit with trees? Christopher Stone, in a famous 1972 article, "Should Trees Have Standing?", makes the point that rights claims continually expand towards new beneficiaries. The reason we start talking about rights for trees is because rights is such a dominant language of ethics. That might or might not be a good thing. There are debates about how rights shut out other ways of thinking and how rights have served capitalistic, industrial, human-centred anthropocentric interests.

The "rights for trees" language is a way of saying we need to carve out a space in which we recognise the inherent ethical significance of these amazing living things. Rights are one tool. They're one particular way of expressing relationships of justice and of injustice. So it makes perfect sense that that's where people go, and I don't think that it's totally implausible to use rights as those kinds of tools in law. I just think there are questions about whether rights are the best way forward.

AN ANIMIST ZEITGEIST

Something is altering in how we look at nature. A shift in attitudes is happening. The nature writer Robert Macfarlane calls some of this movement a "new animism".

But it is not only those who sit in the trees that have this animist sense. I find it all over the place, this sense of reverence and respect for life in all its varied, complex and interconnected forms.

7

PINING AND LOSING

WHEN A TREE IS GONE

There was only a quiet rain when they were dying;
They must have heard the sparrows flying,
And the small creeping creatures in the earth where they were lying—
But I, all day, I heard an angel crying:
'Hurt not the trees.'

CHARLOTTE MEW, 'The Trees Are Down'

The loss of a tree can leave us heartbroken. When it is chopped down it can be a savage blow. When it fades and withers through disease it can make us fret and ache. Ash dieback feels like some slow march of death across the country, against which we are helpless. Even having to move home and leave a tree behind can leave us pining, homesick, filled with both longing and grief.

It's a sign of our attachment to trees that we feel all these things. A year ago, my mum called me, but was unable to speak. My heart lurched. I thought perhaps someone had died. But, in fact, to my relief, the problem was a huge laurel that had been chopped back by a neighbour at their holiday home.

For some people I spoke to the loss of a tree was a significant driver to activism. Paul Powlesland still mourns the loss of a childhood tree, the old oak which stood in the back garden of his grandmother's suburban home. "It was," he recalls, "a massive pollarded oak from the days before the house was built, possibly from the time of Windsor Great Park, five hundred years ago."

His grandmother had been due to sell the house to a developer who planned to demolish it, and the young Paul was shocked to learn that this developer had said they would need to take the tree out before they put in their planning application, as otherwise it would fail.

"I think," he says, "I was thirteen or fourteen and even then I knew it was wrong. I was actually thinking about putting in an application for a tree preservation order. But I felt my family would go absolutely nuts with me and I'd ruin the sale and everyone would be really angry with me and never forgive me. So I didn't and then the tree got chopped down. It was lost. And it's one of my biggest regrets. I wish I'd had more courage to save that tree. It feels like maybe I'm trying to compensate as a lawyer now, fighting for trees. I wonder whether it will ever be enough, though?"

RAGING AT THE TREES

But a tree doesn't have to be chopped down, and an ecosystem doesn't have to be devastated, for us to pine or grieve for it. Sometimes simply the loss of it from our lives, because we move away, leaving it like an old friend, triggers intense feelings. We become treesick in the way we might be homesick, or even lovesick. We find ourselves pining.

Of course, not everyone loves the tree that we love – and this can lead to conflict. In the words of William Blake, "The tree which moves some to tears of joy is in the eyes of others only a green thing which stands in the way . . ."

As we know from the many news stories about neighbourly disputes, a tree can torment a person, send them mad with rage. What is one person's privacy is another person's light obstruction.

One friend even told me of a tree that casts its shadow over her garden. "You think, 'What fool planted that?'" my friend raged. "It's a monster. And every year it gets bigger. It's home to a family of blackbirds who sing at night, but still I hate it."

RICHARD HOLLOWAY
Former bishop of Edinburgh and author

My American father-in-law was worried about the fact that we lived right in the centre of Edinburgh in Jeffrey Street, just above Waverley station. Though we loved it, it was a very urban environment. So he gave us enough money to put down a deposit for a mortgage and invited us to search for a cottage somewhere in the country, where we could go for holidays and the occasional break.

We started house-hunting and found this 18th-century cottage called Pathend – itself a lovely name – near Muckhart Mill, a redundant old mill used as a holiday centre for children. The cottage was down a lovely county lane halfway between Dollar and Yetts o' Muckhart in the shelter of the Ochil Hills – more lovely names. When we saw the cottage in its tumble-down state, we fell for it.

There wasn't much interest in it and we got it. We employed a local architect to modernise it a bit. The cottage faced the River Devon, so it was the back door you met first as you came down the lane. And standing guard at the door was an enormous cypress tree, which I immediately fell in love with. We loved Pathend for the ten years we owned it, and spent all our summer holidays there. We would go up occasionally for a few days and walk with the kids. I would sometimes go up there on my own if I wanted to write. And the parish used it, too. It became a place of rest and healing for us all of us.

Then in 1980, I was invited to move to Boston in the States, so we had to sell Pathend. We spent our last summer there in 1980 before moving to Boston in August. The day we drove

to the airport for our flight, we were quiet in the car as we drove away, till my son Mark, the youngest of our three children, sobbed loudly and brought us all to tears. It was goodbye to Pathend.

We were away from Scotland for six years and didn't come back to Edinburgh till 1986. One day we decided to pay a sentimental visit to Pathend. I was eager to see my old friend the cypress tree again, but I was devastated to discover it had been cut down. Memories of past summers flooded back. I would take the kids for walks during those long days, and as we came back down the field above the cottage, I would play a game with them. There's a lovely cottage down there, I'd point. Look at that wonderful tree. Let's see if there's a kind lady there who might give us tea and scones and homemade

raspberry jam. We'd troop in past the big cypress and find it all laid out on the kitchen table.

Now the tree was gone. It was like the death of a friend no one had told you about. But the loss was not absolute.

An artist friend of ours had used the cottage a lot. Her name was Birgitte Hendil, and she painted birds. But she did more than paint them. She had a mysterious relationship with them, seemed to understand them, and they her. She too had loved the cypress, so she painted it for us, filled with birds, a vivid miniature. We still have it. So, in a way we still have the tree. A great, green flame, alive with birds and memories.

HELEN GRAHAM
Writer and musician

I grew up in a small Cotswold village in the same house my father lived in as a child, which his father had built in the early 1930s. My grandfather, who I never met as he died in his forties, also planted a lot of trees in the large garden. When I was a child my dad often told me his father always used to say he was planting the trees for his grandchildren to enjoy. That idea made a huge impression on me and I loved those trees with a passion. I loved climbing the apple trees in the orchard and eating the fruit, swinging on the branches of the swaying larch tree and harvesting hazelnuts in a dark little grove of nut trees. Most of all I loved scrambling up to sit on the wide low branch of a grand old silver birch tree near the house that was already quite a bit taller than the house.

My sister and I called it the "Horsey Tree" and we used to pretend we were riding a wild horse when we sat up there.

It was the idea of selling those trees to someone else that broke me when my parents divorced and began talking about selling the house. I could have lived in a different house, by all means, but my grandfather had planted those trees and I just couldn't imagine leaving them. So, at the age of twelve, I staged a tree protest! I put some planks up into a big old apple tree and made a platform, then I sat up there all day with some food and my guitar, singing protest songs. I don't know whether my protest had anything to do with it, but in the end the house wasn't sold until shortly after I left home.

JACKIE KAY
Poet

INDIA KNIGHT
Sunday Times columnist

A cherry blossom was the first tree that was important to me because my mum brought me up with this story that she had planted it when she adopted my brother, Max. It was a beautiful tree and I was always a bit jealous that he'd had a tree planted, but I hadn't – it used to really irritate me. I would say, "And what tree did you plant for me?" But this cherry blossom's roots started to progress towards the house and to almost lift the house up, and so it had to be cut right down. It had to be felled because the roots had to be taken up. I was living at home when the tree was felled. I just wanted to weep. It's a great sadness, the loss of a tree and I felt it very deeply.

It is a truly terrible thing to watch something so mighty being reduced in this way. Our tree wasn't ancient — it was like all the big trees that you see driving around: a handsome, friendly oak, about 2.5 feet wide and about one hundred and fifty years old.

Our tree's death warrant was issued a year and half ago, but we'd refused to believe something so beautifully and extravagantly alive, so generous and hospitable to birds and creatures, could be responsible for the sinister cracks that had appeared on the outside of the house. Second, third, fourth opinions were sought: everybody agreed that it was the tree. And so, finally, the tree surgeons came. They expressed sorrow for the tree and talked to us kindly, like doctors or priests. And then they tooled up and started hacking off its limbs.

If you'd told me ten years ago that I would have tears in my eyes about a tree coming down, I wouldn't have believed you. But it is devastating to lose a tree like this – bad enough in a storm, but at least storms are natural events, and the whole thing comes down at once. The night before, I had kissed the tree and said sorry.

ALYCIA PIRMOHAMED
Poet

On my walk to Calton Hill, I spend time identifying the trees. The silver birch in Alberta and Edinburgh are not exactly the same, just as I know my body in Alberta and Edinburgh is not exactly the same.

A landscape alters a body, alters an "I", alters a subject – her imagination.

Calton Hill is so green right now. Seeing the stalks of birch, the wisdom in each dark eye, makes me feel homesick: for the winter birch packed neatly on boreal landscapes between Calgary and Jasper; for those other trees that burgeon vibrantly in Mtwara's red clay. For all the spaces I have treasured.

GWEN WILKIE
Landscape architect

"The Moories" was the name that we gave to the mysterious people that we imagined lived in the ghostly winter larch – an army of eerily undead trees. I was a bit older when my dad would routinely inform me: "An interesting fact – larch is the only deciduous conifer." It was not strictly true, but what did I know?

Larch is not native to the UK, but it is a fantastic forestry tree – fast-growing and adaptable, with strong and durable timber. Moreover, its shifting quality, its changing colour and texture, is often a crucial counterbalance to a blanket of evergreen. It plays a significant role in establishing forests that sit well within the landscape. Forest design is a fundamental part of my job as a landscape architect in Forestry and Land Scotland. This is why, two years ago, when I started this job, I was absolutely horrified at realising the scale and impact of the disease that currently has a stranglehold on our larch, Phytophthora ramorum.

P. ramorum is a fungal-like pathogen that affects many species. Larch is particularly susceptible. It is spread on wind and rain, but it also travels by foot, carried unwittingly by people. In Scotland the disease has so far been particularly cruel to the wet and windy west coast, arriving unplanned and unwelcome, and giving foresters a challenge like nothing they have ever encountered.

Will larch still be around in the future? There will certainly be less of it in my lifetime. We are working very hard to make our forests resilient and adaptable places, not just in response to tree disease but in response to climate change. I just hope the Moories are adaptable too.

A GRIEF THAT'S BIGGER THAN A TREE

But a tree is often not just a tree. In these times of biodiversity loss and eco-grief, tree loss can seem part of a wider guilt and grief over what we humans have done to the natural world which is our home. Extinction Rebellion co-founder Simon Bramwell, says the loss of the trees he fought for, put him in contact with what he describes as an "omni grief" for the natural world. "These days psychotherapists might call it solastalgia," he says. "The loss of our environment, and the grief we feel for that."

Bramwell recalls that when he was spending weeks in the depth of winter, trying to prevent the trees at Stapleton Road being felled for a bus lane, developing a relationship with the tree he inhabited, he and his companions went through a powerful grief journey. Whether for just one tree or a global ecosystem, grief feelings can be intense. Simon says, "Sometimes grief drives us mad, and sometimes it brings us closer into the embrace of life."

8

TREE SPIRIT

THE SACRED GROVE AND THE FAIRY KINGDOM

*Praise and blame, gain and loss, pleasure and sorrow come and go like the wind.
To be happy, rest like a giant tree in the midst of them all.*

BUDDHA

When we stand in trees, we can often feel there is something there that is more than physical. It's hard to define, and not quite graspable. The celebrant Tim Maguire describes that feeling as numinous:

"In the same way as we feel in front of great art, we have a connection with nature we don't entirely understand. We feel something. It's not a particularly reasoned response, but it's a spiritual or emotional one."

We have a long history of seeing something spiritual in our woodland glades and forests. They are central to both polytheistic and monotheistic belief systems. Sprites and faeries and dryads live there. In Scottish folklore the forest is home to the Gille Dubh, a gentle, well-meaning guardian fairy, who, it's said, will guide any child home. The Buddha was born under the bodhi tree. Norse mythology is based around Yggdrasil, an immense world tree, as the centre of its cosmos. Adam and Eve ate an apple from the tree of knowledge. Hindus pace around the sacred fig tree. And here, in the UK, trees are central to Celtic spirituality. Even the word "druid" means those with the knowledge of the oak.

Some of those I spoke to talked of how their love of trees connected with a monotheistic faith. Some told me of their paganism. But many talked of a loosely spiritual feeling, or a worldview that revolves around reverence and respect for the natural world of which trees are a key element. Even people who don't really regard themselves as religious will tell you that they feel there is something sacred in these wooded settings.

Tim has conducted many weddings in what he calls sacred groves, and observes that they have a long tradition. "The most obvious one, in the western tradition, would be the sacred grove of Dodona. Homer talks about it. There are lots of sacred groves in history and they exist in the druid past and the Celtic past and the Germanic past as well."

As Tim points out, Christianity had a liking for building on the sites of these groves - and, through the naves of its churches, even creating something that looks remarkably like a tree canopy, of branches stretching out towards each other, in stone.

HUDA

Participant in the Branching Out programme

I find nature connection a way to get closer to religion. Appreciating and admiring nature can make you feel closer to God. It puts things into perspective of how our universe was created and how we are something so small in a forest of imagination and creation. Every small creature makes an impact, every tiny plant has its purpose.

JOANNA SORAGHAN
Community cycling manager

CONAILL SORAGHAN
Renewable energy engineer

It was in my head that if we ever got married I would want to do it at this spot at the Hermitage, Dunkeld. It's a place that we love and would visit whenever we were travelling up north. It's really famous, it's got trees that were the tallest in Britain, and some really old trees. When I think of the trees there, it's an area they call the cathedral that I think of. You're standing there and you look up and you feel so tiny I loved standing there.

You can't doubt the healing power that comes with trees. I definitely believe there is something that connects humans to nature that you don't get if you're in cities all day long. You need to be around it. But neither of us are explicit about this, it's just implicit, that we're always outside, always running. We understand the value of nature but don't shout about it or preach about it.

When we decided to get married, we did a recce and scoped it out. There's a nice long walk into Ossian's Hall, and neither of us wanted it in the building – we wanted it to be in nature. There's a clearing they call the cathedral and then you come to the waterfall. The waterfall is a drop down and we were both looking at it and it felt really right. It's right beside the waterfall, which is quite loud. But there was this perfect clearing. For us it was about being outside in nature, about bringing the two clans together.

YU ZHANG
Student

I particularly like old big banyan trees. When I was a kid, I felt they had magic and could talk when no one is around. I always wanted to be a tree in my next life, so that I could stand in a forest and see how the world changes.

RABBI NATAN LEVY
Interfaith leader

Planting a tree is about hope for the future – you don't plant these trees for yourself, you plant them for your children.

There's an allegorical Jewish story, in the Midrash, about a man who plants a tree and falls asleep and wakes up seventy years later and gets to see the fruit of what he has planted for his children. It touches on the theme that we plant these trees now but won't see their real fruits.

———————————

Tree-loving is there, right at the heart of our native paganism too. Liz Harris, an ovate who is also a practitioner of the Celtic Ogham, speaks about the role of trees in her practice and beliefs.

LIZ HARRIS
Druid

As Druids, we work closely with the trees. We revere all life, but especially trees. The word "druid" means, "Those who possess the knowledge of the oak", and indeed the oak tree is symbolic of Druidry.

Trees are living beings; without them we would not exist. We breathe in the beautiful clean air they produce and, in turn, the trees take our toxins released on our out breath. The perfect symbiotic relationship.

In Druidry, we live by the seasons of the year. We honour eight festivals in the year, all celebrating different seasonal changes, that reflect the changes in our own lives as we walk the wheel of the turning year. Druids celebrate together in groves, whether this is outside in Mother Nature, as the elements provide for us, or connecting with our inner sanctuary, our inner grove, as we journey within ourselves.

The beauty of this is you can do this at any time – seek guidance from the trees, from our guides and beyond, imagine in your inner mind sitting in a grove of trees . . . in peace and tranquillity, connecting at a deep level with Mother Earth, nature and all life. You can feel the energies building as you sit in meditation, and after a bit of practice, this can be your personal sanctuary.

I am also an Ogham Tree Therapist and Celtic Reiki Master. The Ogham is the ancient Celtic tree language of the Celts and Druids, often used thousands of years ago as funerary inscriptions, or markers of boundaries on land.

I work with the energy signatures of the trees in Celtic Reiki and the essences produced from the twenty Ogham Trees in Ogham Therapy. I see today so many people with "dis-ease" in their being and all this is disconnection from their core, from their truth and Mother Earth. When we allow the flow of energy from source, we activate our true potential and our soul purpose.

There is a horse chestnut I have been working with for over a year. I've seen her splendour in all seasons of the year. The horse chestnut is not in the Ogham Trees, but that doesn't matter. I always ask permission of the tree before I work with it, this is only respectful to do so, and as long as you ask from the heart, then listen with all your senses, all will be as it should be.

There are many ways to connect with trees, and each person has their own unique way, whether it is by touching the bark, the leaves. There are no hard and fast rules. I usually sit with my back to the tree, as close as I can, and place one hand behind me on the bark, the other on my solar plexus, both level if I can . . . this way I can sense the flow of energy from the tree.

In a meditative state it is possible to feel the energy signature from the tree, even the heartbeat, which is rhythmic and slow. Allow yourself plenty of time, never rush these things or you may miss so much. Listen with every part of your being, and be open.

The horse chestnut for me symbolises

longevity, to bend and shape, to adapt to change, to be in free flow. But if you sit with this tree, you may get an entirely different message and that is just as it should be. For example, the signature of the oak is strength and courage. The alder is protection. The birch is new beginnings and initiation. The pine or spruce are far reaching and seeing beyond the limits of illusion.

Our ancestors honoured trees and that is how we should be working with them today, as they have so much to teach us.

TOR WEBSTER
Tour guide

The holy thorn tree in Glastonbury has inspired me greatly and is part of my day-to-day work. I start every tour from the site at which the significant holy thorn tree was planted on top of Wearyall in Glastonbury. Sadly that was vandalised in 2010 – someone drove up in Land Rover and tried to rip it out of the ground – and more recently the rotten trunk was destroyed.

It was a unique tree – it not only flowers in the springtime, but also at the winter solstice. It's not indigenous to these lands. The story tells of Joseph of Arimathea coming and planting his staff there.

In 2010, I was in New Zealand, travelling, and I fell ill with a really high fever. I was knocked out in bed and I could feel that there

was something else going on. And then I turned on social media and I realised the time that I got ill was when the tree was destroyed. I do believe that when you build this connection with a special place that you are linked in some way.

One of the things that has made me positive, since the holy thorn was destroyed, is that, at the gateway to the tor there were two oak trees planted to honour the earth mysteries expert, Anthony Roberts. Just after the millennium, at fifty years old, he walked up to the top of the Tor and never came down. He had a heart attack at the top and had to be airlifted off. His family planted two oaks at the gateway to the Tor.

These two oak trees create a gateway back into town. I've called them Arthur and Guinevere because Guinevere is quite sheltered by the other trees around her and it's very difficult for her to get her leaves because she's so sheltered, so Arthur who is very much in tune with her, waits for her.

When I take people on a tour I always suggest that they pause for a moment under these trees. And what I do when I come through that gateway after coming off the tor is to stop and put my arms out to both the oak trees and I feel the energy coming through my hands into my heart and coming up and then I see oak leaves coming out of my mouth like I am turning into the Green Man. For a moment, I become the Green Man.

WHERE FAIRIES DWELL

But you can't talk about the spirituality of trees and not mention the fairies, the mischievous and protective spirits who have long been thought to inhabit our woods. Whatever you think of them, the little people must be acknowledged, as they were by many I chatted with.

Louise Gray, author of *The Ethical Carnivore*, for instance, says she feel that certain trees "are where the fairies are". "There's one I visit in Wester Ross in the northwest Highlands. It's magic. It's out of control. When I'm there, I generally fall in a bog or something and I feel like there's something bigger than me playing tricks. I feel like I'm being watched."

Storyteller Amanda Edmiston recalls that as a child she would often go up to the trees to ask them questions and wait to see how the branches waved. She observes that we have, across the world, lots of stories about the spirits of trees being embodied in a creature that we then call, say, a fairy or a dryad.

"I think, underneath it all," she says, "this is a way of expressing the sentience that trees have. They clearly do have a personality. It's just different to what a human personality is and we don't necessarily have a way to articulate it that feels acceptable. It doesn't feel scientific and grown up enough for modern society."

We now, she observes, know that trees can talk to each other through root structures and mycelium. That's part of our story about who and what trees are. "Maybe," she postulates, "turning them into these fairies, or magical, mystical beings, was the acceptable way of doing that in times gone by, just a different era of explanation. It's about looking at important and valuable and very beautiful beings that we connect to and at our relationship with them."

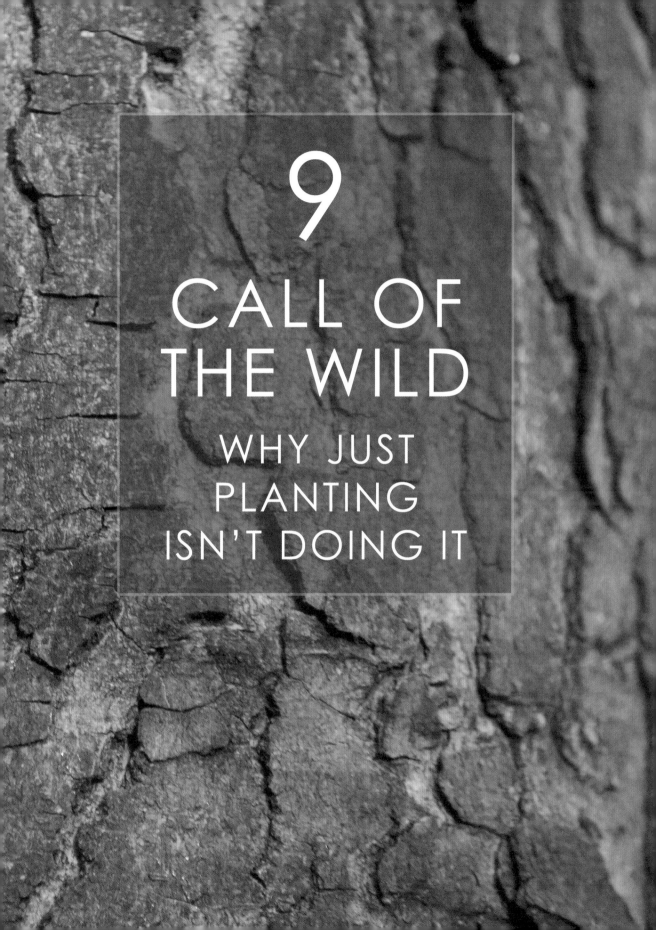

9
CALL OF THE WILD

WHY JUST PLANTING ISN'T DOING IT

Whirl up, sea—
whirl your pointed pines,
splash your great pines
on our rocks,
hurl your green over us,
cover us with your pools of fir.

H. D., 'Oread'

The bare, high mountains of the Scottish Highlands can give goose bumps of wonder, but many of us no longer look at them as romantically wild and untamed. We now know that they lack something. As forest ecologist, Kenny Kortland, puts it they are "an anthropogenic landscape", crafted by thousands of years of human impact, "ecological deserts compared to what nature would create".

Biodiversity there is so low, he says, that any kind of plantation, of any kind of tree, even a stand of the much-derided non-native Sitka spruce, would be better than what we have now.

One of the people who has done most to raise awareness of the fact that the bare, biodiversity-poor uplands of the UK are manmade is the environmental writer, George Monbiot. In his rewilding book, *Feral*, he described a similar ecological desert in Wales, where he then lived. "Sheep farming in this country," he wrote, "is a slow-burning ecological disaster."

Monbiot offers an account of how conservation had frozen ecosystems in time, preventing the reversion of heath and woodland to moorland, and proposes, instead, a different approach: rewilding, a process "driven not by human management but by natural processes".

As we rage about the burning of the Amazon rainforest, we might remember that the UK has one of the lowest levels of woodland cover in Europe – 13 per cent of forest cover compared to 39 per cent in France and 74 per cent in Finland.

The plan, of course, is to increase it. The Woodland Trust has a target of raising levels from that current 13 per cent to 19 per cent by 2050. Friends of the Earth want it doubled. But a key issue and debate is how to make sure the right trees are planted in the right way, or even whether it is a matter of deliberately planting. Some believe that more important than planting trees is allowing mature forests to spread. And that, of course, is rewilding.

The idea is gathering popularity, along with evidence of its success. The word is now so popular we talk of rewilding our gardens, rewilding ourselves. But more importantly, there are now major projects across the UK in the process of reforesting, rewilding or regenerating. In Sussex, there's the pioneering Knepp estate, run by Isabella Tree and Charlie Burrell. In the Highlands, there's Cairngorms Connect, where estates such as Glenfeshie and RSPB Abernethy are reducing deer and seeing the forest return. The tree is coming back, and, in these places, without a plastic sapling tube in sight.

ISABELLA TREE
Rewilder and author of *Wilding*

Our rewilding journey started when we got Ted Green to come and look at the Knepp Oak. We had been worried about it because it's such a huge tree and it was beginning to split down the middle. During the Second World War, when the Canadian army had been stationed at the castle, they had tied it together with tank chains, but that was beginning to fail and we

were wondering what we could do to save it. Someone told us that Ted, who was advisor to the Crown on the great oaks at Windsor, was the man to talk to.

But when he came, we found he wasn't concerned about this tree at all. Actually it does look in the bloom of youth. If you look at it from a distance, it doesn't look like an old tree. It's just thriving.

He advised us that we could cut the canopy back a little, over a series of years, to decrease the wind blow and stabilise it a bit more, and we put in a few wires, judicially. He wasn't really concerned about that tree, but when he looked around at the other oaks in the landscape, he said, "They're in trouble."

We hadn't really thought about the other trees in the park. They're much younger than the Knepp Oak, which we think is about five hundred years old. But they look much older because they've got these dead, staggy limbs, now, even though they're mostly around two hundred to three hundred years old.

The land they are in was ploughed up in the Second World War as part of the Dig for Victory campaign. That ploughing had begun to affect those oaks, and also, in the years that followed, the chemical fertilisers, herbicides and pesticides, too. But the pace of a tree is so different to the pace of a human. We hardly noticed they were beginning to grow quite staggy by the 1990s, it was so surreptitious.

This moment with Ted was an epiphany.

We started to see trees as part of something much greater than simply individuals. Now we realised a tree is just the tip of an iceberg. Underneath it's connected to this huge invisible world of the soil – something we're only just beginning to understand and we've neglected for decades. Everything comes down to the soil. So now when I look at a tree, instantly I'll think of all the mycorrhizal root connections and the other trees, shrubs and vegetation around it that it might be talking to.

That's what Ted taught us. He rather provocatively says, "What is a tree actually, but a fungus? It's a giant fungus." The mycelium goes all the way into the tissues of that tree. What you're seeing is mostly fungus.

We learned also that we were too keen to get rid of dead trees. We heard Ted talk about oaks and how they support life, and how they age, and how they hollow out and become roosts for bats and birds, and then you get the guano, of course, from all those creatures, giving extra fertiliser to the ancient tree.

In the early days of our rewilding project, changing that mindset was one of our key challenges. We suddenly thought, we've been tidying up all our lives. When the next oak died right in the sightline of the house, Charlie, my husband, was itching to get out there with a chainsaw and chop it down. Yet the new conservationist in us was saying, "No, we're going to leave it." And just leaving it right there in our view, waking up to it every morning, meant we began to look at it completely differently. We started seeing sparrowhawks nesting in it, and a fox digging for short-tailed voles. Herons perched on its branches for hours. You suddenly realise there's a whole universe happening around that tree.

NATALIE TAYLOR

Artist

We own a small mixed woodland in Fife, which is a huge privilege and joy. We feel quite responsible for it in so many ways, as trees and forests have locked within them so many keys to future human survival. Ours is partly a handed-down spruce plantation, about twenty years old, and partly a mixed young woodland of oak, silver birch, Douglas fir, Scots pine and hawthorn. We've also planted over two hundred saplings, whips and fruit trees to extend the biodiversity as much as possible.

CAMERON McNEISH
Mountaineer

I've always loved Scotland's mountains and I used to think this was what fine landscape was all about – bare slopes, this rugged beauty – but in recent years my feelings have changed.

Frank Fraser Darling, the noted naturalist, once described the Scottish Highlands' landscape as "a wet desert," a bare landscape lacking in vegetation and particularly trees. We have lost most of our native trees because of an over-enthusiasm for sheep and sporting estates, where the deer eat the young trees before they even get a chance to grow.

When you climb a hill your ascent often begins in a forest, it's part of the mountain experience, indeed pine forests often form the skirts of the mountains. At the end of the hill day, as you descend back into the forest and perhaps feeling a bit weary it's almost as though the air has extra oxygen in it. It always gives you a bit of a boost. You smell that lovely pine resin and the scents of the forest which are quite different from those of the bare mountain. It revitalises you.

What we're now learning is that in Scotland trees could grow much higher up the hill than they have been. Not long ago, I came across a pine tree just off the summit of Schiehallion, which is 1,083 metres above sea level – this single tiny pine tree, peeking through the screes and rocks of the summit ridge. This is what is happening there because sheep have been removed from the hill and deer numbers drastically reduced.

I remember a conversation I once had with the noted naturalist and conservationist, the late Dick Balharry. We were going for a walk near the Corrieyarack Pass, that historic route that runs from Laggan over the hills to Fort Augustus. I made the comment that the landscape looked terrific, but Dick didn't share my enthusiasm. "Don't you realise that's an industrial landscape we're looking at? All the trees have been removed. Between here and Fort William there are miles and miles of underground tunnels, and shepherds have been brought in here to have their flocks of sheep keep the grass short so that there's a good water run-off. The water runs off into all these tunnels, and goes through to Fort William to feed the aluminium works at Fort William."

I had no idea. He explained to me that this area should be one of woodland with trees – birch, rowan and aspen – growing up to between 400 and 600 metres on these slopes.

It was Dick who talked Scottish Natural Heritage into reducing deer numbers throughout the Scottish Highlands. At one point in his career he was in charge of an area that included a mountain called Creag Meagaidh, midway between Ben Nevis and Cairn Gorm. Birch woods covered the lower slopes of Coire Ardair, but they were dying because there was no progeny. The place was overrun by sheep and deer. No young trees were getting through, they were being nibbled at birth. Dick encouraged his employer, SNH, to remove the sheep

and reduce deer numbers. No fences, just stock reduction. In the twenty years since, the regeneration in the woodlands has been absolutely incredible.

The success of that experiment at Creag Meagaidh has inspired others, including Anders Holch Povlsen, the Danish clothing billionaire, in Glen Feshie and other Scottish estates he now owns. In the time since Povlsen took over, the difference in Glen Feshie has been astonishing, the rate of natural growth has been quite amazing and wildlife has been attracted back to the area. And much of that rewilding really began because of a local chap by the name of Thomas MacDonell.

Thomas was born in Kincraig at the foot of Glen Feshie. He had a fencing company with his brother and they did a lot of work for Scottish Natural Heritage. As they were doing these fencing jobs, Thomas noticed something odd.

Inside the area that was fenced, luxurious undergrowth and trees were predominant, and outside the fence, almost looming up against it,

desperate to get in, lay the carcasses of starving deer. That made him think and he started researching and talking to people, including Dick Balharry, who said that this is what happens when you build fences. You create islands of fertility – and outside it's still a wet desert and there's not enough food for the deer. Thomas's wife, Ali, was working as the chef at Glenfeshie estate at the time so Thomas began doing his own little experiments.

When Povlsen bought the estate, Thomas said, "Look, would you mind going out for a ten-minute drive in the Land Rover, I want to show you a few things?" And in that drive he encouraged Povlsen to rewild the place. Povlsen was convinced and today Thomas is the conservation director for Wildland Ltd, which is Povlsen's rewilding company.

We owe people like Dick Balharry and Thomas MacDonell our thanks for their vision and for their commitment to a better and more naturally diverse Scotland, a Scotland with more trees.

KENNY KORTLAND
Forest ecologist

Just about any kind of manmade forest is valuable for wildlife – even the much-derided plantations are amazing places for biodiversity. In Britain and Ireland, plantations have almost equivalent levels of species-richness for certain taxa as native woods, yet they have previously been described as monoculture deserts. They're very biodiverse – and this is peer-reviewed, published science. We should all be celebrating that – because we all consume timber, we all use toilet roll and pencils, and have kitchens made of wood.

One of the things that is remarkable about the Cairngorms Connect project area is that it is an island without illegal persecution in a landscape of predator eradication. There's lots of new research coming out that if you have the full complement of predators, they exert top-down influences on their ecosystem, known as trophic cascades. The structure of wildlife communities is determined partly by the downward pressure from the predators. For example, in Yellowstone Park, when the wolves came back, they created a "landscape of fear" for the deer, habitats were allowed to thrive again and biodiversity went through the roof.

We don't have the top predators in Scotland. We don't have lynx or wolves. But we see a reassembling of the predator community here that is really fascinating. We've got ten or eleven breeding species of avian predators, we've got golden eagles at almost saturation density in the Cairngorms Connect area, we have white-tailed eagles, hen harriers breeding, goshawks breeding in these forests for the first time. And, from first-hand experience of working with Police Scotland, I know goshawks are extremely heavily persecuted elsewhere.

The prevalent psychology in Scotland is: "Predators are a problem – we've got to get rid of them." So, when the goshawks returned a lot of people said, well, that's the end of your capercaillies. But what we're finding is goshawks eat birds like jays, which have returned to the forest as well, and jays are a key nest predator of capercaillie, so the goshawks are actually doing the capercaillies a favour. We're getting a more diverse and robust ecology.

There's various research that shows there's plenty of room for lynx in here. If you're worried about mesopredators like foxes, the best thing to do is to bring in lynx and they'll sort out the foxes. However, it's also about how people feel and there's no doubt that lynx would eat some sheep. The ideal system would include the lynx and it would include wolves.

HAMZA YASSIN

Wildlife cameraman

I was born and bred in Sudan, North Africa, and there we don't have many trees. I moved to the UK when I was eight. Coming here you could immediately see the greenery. In Sudan once summer hits everything goes yellow and when it rains everything goes green. All we could think of was that England is green. *Why is it green?* We were expecting snow.

Just behind where I live on the Ardnamurchan peninsula, is a very old, ancient forest of Scots pine and larch. Seven or so years ago, before I came up here, I'd read that there was a golden eagle living on a larch here. Then, about five years ago, I found this very old remnant of the nest. I thought, Oh my God I've found the tree this book described. Then, a year later, I was doing my regular walk of the area and I thought I'd go and check out this larch tree. I started to climb up towards the nest, and then I heard the *cu-cu-cu-cu* noise.

I thought, I know that sound. It was a white-tailed eagle nest. A pair had taken over the golden eagle nest and built it up. The nest is about the size of a double-bed mattress. This year I got a licence to film them.

I tend to find all the animals I need – the Scottish big five of the red deer, the golden eagle, the white-tailed eagle, the capercaillie, the pine marten – near trees. Trees are such an important part of life. People talk about regeneration of the land, but there is a certain amount of wilderness that we will never achieve in our lifetime. It's great that we are replanting, and we need to replant as much as possible, all broadleaf plantations, not conifer plantations, but these big, old trees that are able to house golden eagles and whitetail eagles take a long time to grow. A tree like the one they've nested in is between 150 and 200 years old. You can't just plant one of them today.

Hamza Yassin

SHAILA RAO
Ecologist, Mar Lodge

We were probably the first estate to adopt a zero-tolerance approach to deer management. In the early years we had target figures for deer numbers. But it was always difficult to know how many deer were on the ground at any one time – when the stalking team should shoot deer and when they shouldn't. When we made the decision to have a zero-tolerance approach, it became much easier in practical terms. Any time the stalkers saw deer, they would attempt to get rid of them.

It seems brutal, but it was required to relieve browsing pressure and give the young trees a chance to grow. But the reality of the situation is, even with that severe approach, in terrain like Mar Lodge Estate you never have zero deer. There's always some deer that remain.

The idea of culling rather than fencing, was partly motivated by the fact that it's quite impractical to fence the land here.

Deer are beautiful animals and Scotland has thrived on the emblem of a stag. But there's nothing out there managing the deer population or moving them around in the way that wolves once did. And the evidence suggests it's partly predators moving the deer on which means they don't get the opportunity to run down one place and graze it thereby heavily suppressing tree and plant growth.

When we started the culling, there were a lot of people against what we were doing here. We shot a lot of deer – large numbers on an annual basis for a number of years – and at first it seemed nothing was happening, as if the forest wasn't regenerating.

There were a few years where the NTS themselves were questioning whether this was the right approach, and we were getting a lot of criticism from our neighbours, the local community and other people.

But the reality was that there's a bit of a lag between reducing the browsing pressure and visibly starting to see the trees actually grow and appear. It takes two to three years of lower deer numbers and reduced browsing pressure before there's obvious apparent change in the landscape. A similar thing was experienced in Glenfeshie, and Abernethy. There was a lag in terms of the response of the vegetation to the culling efforts.

What's quite fascinating now is that the local community, and even our neighbours, are more accepting of what we're doing now that they can see the trees. It becomes more obvious to them now, the impact that the deer were having. When people walked around here twenty years ago, they didn't see any young trees but didn't really notice that. But now they see all the trees coming and it dawns on them that it was deer browsing that was stopping that happening.

The problem is people want to see change in their lifetime. They like to feel they're having an effect. And the hardest thing to do in conservation is to sit back and actually do nothing. That's the most difficult thing to do.

We're really seeing quite a dramatic change now. The first few years were slow, but now it's rapidly changing. It was definitely about holding on. And I think the Trust came close to not doing that because they came under so much criticism. But thankfully we held firm.

LET'S CO-CREATE

One of the biggest questions around rewilding is how you do it in a country with so little woodland cover, where there are few trees to self-seed and start the process. The forest restoration pioneer Alan Watson-Featherstone, founder of Trees For Life, has pointed out that letting the wild come back isn't going to get us very far given that so much of our land is now treeless thanks to our practice of keeping large herbivores on it. So, yes, we do need to plant. As Alan has said, in a TedX talk, we need to "co-create with nature".

And planting, actually, is something we can all get involved in. You don't have to be a landowner with vast acres of moor to get involved in this process – there are ways of doing it as a volunteer. It's even possible to restore or rewild by helping to bring trees back to parts of our cities.

LOUISE GRAY
Author

My family often say we have a tree-planting disorder. Some years ago, with the help of friends and a grant, I planted about ten thousand trees in Wester Ross. People think planting trees is about having something to mark you when you're not there. But what's so good about it is that it dissolves the ego. Perhaps you think you want to plant a tree so it's still growing there when you're gone, but actually what it makes you realise is how irrelevant you are. And it's so good for us, so good for your mental health to get that feeling. It was transformative for me. Planting all those trees made me a more grounded person.

KATIE SMITH
Hospital volunteering coordinator

I got involved with doing things with trees partly out of climate change fear. Because of that I did tree planting with Trees For Life and that made me realise how easy it is to grow a tiny tree and put it somewhere and, provided it doesn't get eaten by the deer and the rabbits, it will grow.

Every day on that week we were either in the little tree nursery or up the hill where Trees For Life have been planting systematically a big area that's fenced off from deer. We were taught what would be a good spot for the trees and we would just bung them in – in a loving way. They were very clear that you should bung them in with love!

Something about spending the whole day doing that physical labour to make the world a better place just took away the anxiety of all the other things that needed to be done. You had that feeling that you were doing everything you could do.

So, I thought, right, okay, I'll just grow some in Edinburgh where I live. I collected a whole load of seed that autumn and started growing trees. I thought, I'll ask other people to look after them and then maybe we would have a tree nursery across the city that would be looked after on lots of people's balconies and back gardens.

Quite a few of my tiny trees haven't made it, but I'm learning. I also applied for the free trees that the Woodland Trust offer to communities – so now I have been involved in helping others plant 1,100 trees across the city.

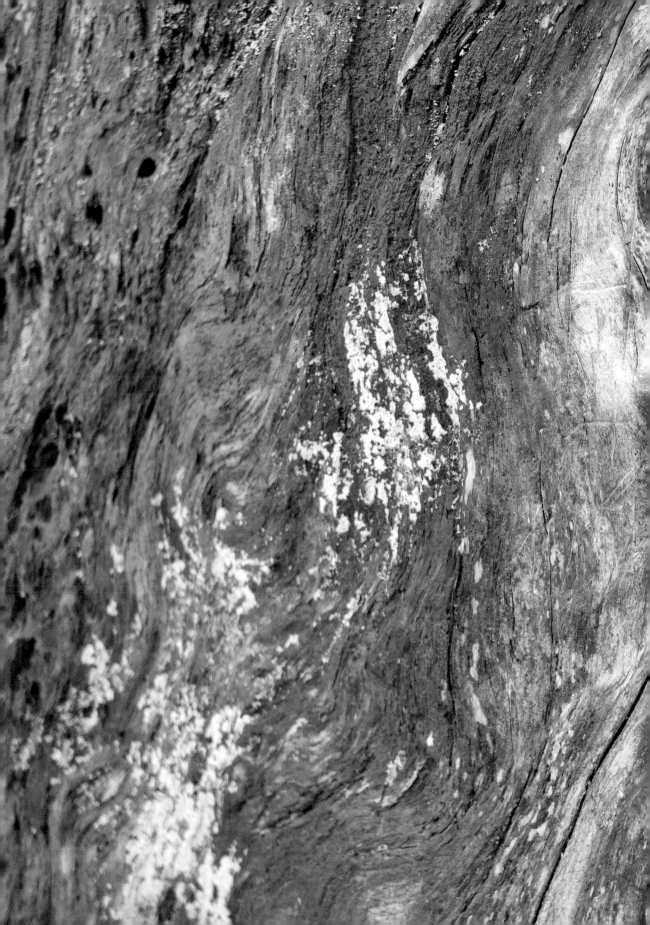

10

THE GIFTS OF
THE TREES

THE ART OF TAKING
ONLY WHAT WE NEED

I had a little nut tree
Nothing would it bear
But a silver nutmeg
And a golden pear . . .

NURSERY RHYME

"What you make from a tree should be at least as miraculous as what you cut down," writes Richard Powers in *The Overstory*. We could fill a book with a list of the gifts we receive from the forest. Oxygen, paper, books, timber, firewood, syrup, apples, pears, cherries, walnuts, hazelnuts, toilet roll, rubber, cork, scents, resins . . . and that's before we get started on the multiple uses of timber, or the gifts from trees that are not material.

Those who know those gifts best are the people who work with trees. They are those who carve them, lathe them, chop them for firewood, press cider from their apples, make tinctures from their leaves or forage for fungi at their base.

Most of us, as herbal storyteller Amanda Edmiston points out, are aware that trees are fundamental to our survival. "They provide vital resources for human life as we know it. Paper, medicine, tannins for dying, ships, everything comes from a tree. Even though it can sometimes seem that in the clamour of the modern world, people have forgotten that, I don't think we have. I think we know it."

The idea that trees have gifts and treasures to offer us, but that we should use them carefully and not take too much is there in a story Amanda tells titled, "The Dancing Trees". In this tale, a young boy, Jack, learns that every now and again, on a midsummer night, the trees rise up out of their root holes and dance, and that when that happens it's possible to go down into the chasms left by their roots and find treasures. Take too much, though, stay too long, and you could be in deep trouble – as one character in the story discovers.

TOM KITCHIN

Chef/Owner, The Kitchin, The Scran & Scallie and The Bonnie Badger

When I went to London as a young chef, I worked for Pierre Koffmann, at the three-star Michelin restaurant La Tante Claire. This was when the restaurant was located on Old Hospital Road and we were closed on Sundays and Mondays. By the time it came to your days off you really needed them after the gruelling long working hours. As Chef Koffmann started to get to know and trust me, he would sometimes invite me mushroom picking with him in the New Forest, which is just outside of London. We would leave really early in the morning, like 5 a.m. and head out looking for mushrooms. It was pretty incredible. Chef taught me about the different kinds of wild mushrooms and when to pick them. It was a great introduction and education for me on foraging and to appreciate and understand the seasons better.

These experiences changed my relationship with the forest. As a youngster, growing up in rural Tayside, I took the countryside for granted, like most kids do. It was only as an adult and a young chef when I began to find the joys of nature and finding something wild and the incredible excitement in preparing, cooking and enjoying it.

I think many Brits are scared of wild mushrooms and picking the wrong ones. Foraging is really enjoyable and I see that when we take our friends and their kids mushroom hunting. We drive to a secret location in the woods and play a little joke on the other family by blindfolding them because we don't want them to know where we are. It's a lot of fun.

We have a whole range of foragers who work for us at the restaurants. For some of the foragers it's their full-time job and some others who do it just because they love it. Generally speaking, the most passionate ones are either Eastern European or Italian and part of their heritage is mushroom picking. They're always amazed by the bounty of Scotland.

My wife, Michaela, has been a big influence on my foraging. She's Swedish and her grandfather, Sven, was a big man of the forest. He would always take Michaela and her brother Fredrik on these journeys, and teach them about foraging and so many aspects of forest life and living in harmony with nature. When we go out mushroom hunting now as a family, my wife goes behind me and picks mushrooms and wild herbs that we've walked over because she's been brought up picking them. One of her words of advice about mushrooms, from Grandfather Sven, is, "If the mushrooms look scary, they probably aren't good for you." My advice to anyone wanting to head out on a mushroom hunt is to always only pick the ones you're 100 per cent sure of and yes, stay away from the scary-looking ones for sure.

ROB AND GABRIELLE CLAMP
Birch sap harvesters

We were novices when we first started tapping birch trees and didn't realise back then that actually it is a fine art. Our first experience of birch tapping was around three years ago, on a very cold, frosty morning. We were standing in a birchwood with trees that were very old and gnarly and full of character and quietly asking their permission before we drilled a small hole through their bark. With birch tapping you never quite know when the sap is rising until you tap a tree and check, so we were unsure as to what might happen. We nervously drilled

a small hole and waited for the sap to appear. After a few seconds the first droplets ran slowly out of the hole. We waited for another minute and the rate increased to a steady drip and at this point we put our wooden tasting cup underneath to collect some. The best way to test if the birch water is ready is to taste it. We had our first taste and were unsure as to what to expect, and were thrilled to find the birch water had a silky texture, was crisp and cold and had a pleasant afternote of cucumber or melon and hint of slight sweetness.

WOODEN TOM
Wood carver

I developed my love of trees while growing up in Birtley, a rough ex-mining town in the north of England. You go back there and it's quite a contrast to where I live now in the Cairngorms.

Trees stopped me from getting into trouble. Plenty of kids just don't have much to do in a lot of towns, and they've got so much energy. If there's nothing for them to do, they'll just get into trouble. I definitely would have gone that way, but getting into the woods saved me.

I work a lot now with birch – carving and teaching carving courses. My earliest memory of being impressed by birch, was I think Ray Mears when he was first on the TV, and he showed how you could tap a birch tree. I saw that and I was blown away.

I started looking for a birch tree and I found a scruffy, littered little patch of birch out the back of a petrol station in the middle of Birtley, on my way to school. I must have been about ten years old. I stuck a knife into it and it started pouring out this clear sap, like water. I tasted it and it was sweet. That little connection was phenomenal.

Birch is the nicest wood to carve. It's softish, but it's hard enough to be durable. I do green-wood working, which is about working with wood while there's still moisture in it, and the wood is much softer, and because it's softer and easier to work with you can use hand tools and that has a knock-on effect as well because you don't need power for them. To make a living I need either two big or about five small-to-medium birch trees in a year – that pays the mortgage, bills, all the holidays. I see my living in terms of those trees.

STEVE BURNETT
Luthier

What I always say is that when the wind starts in the woodland you hear the woods speak as vocal chords of the Earth.

We've lost our connection with Mother Earth. Trees are there, and they're ancient living beings and they pick up all the energy from the soil. I'm a great believer in energy. I take the energy from the trees and transfer it to music.

For many years I've been making violins. That all came about through a heightened awareness of the need to protect trees and our environment. When I was very young I got chronic asthma, which on many occasions resulted in asthmatic attacks. You only have to feel you are unable to breathe, that you're dying from lack of oxygen, to appreciate oxygen, and the trees, the lungs of the Earth.

I can't seem to just make a violin for the sake of making a violin. I need to have an artistic metaphor so it can inspire through the power of nature and music, and through the power of sculpting a bit of wood in a traditional fashion.

What I try to do is use wood from trees brought down in storms, or the branch of a living tree, or driftwood. This unique violin is from driftwood. It was dubbed Il Mare, which is Latin for the sea, and endorsed by Marine Conservation as "a voice from and for the sea" and has played at a number of international Marine conventions.

I've been making violins for thirty years and for over twenty I've been looking for a suitable bit of driftwood. It was walking on Gullane beach [near North Berwick] that I saw a log that had washed up. It's a metaphor in itself. The back and sides are wood from the sea. The wooden linings and components inside the violin are willow from a Union Canal tree brought down in a storm end of 1999. The front is a bit of spruce from the mountains and the neck is sycamore from the Water of Leith. So there's a story here of from the mountains to the rivers to the sea, which is the water cycle.

As far as I can tell, the driftwood came from a poplar tree and it's been out at sea for quite a long time. It's quite a light wood and I like to think it's been drifting about picking up all the kinetic energy of Neptune's world. Who knows where it's come from. It's probably been in a far-off place. It was almost serendipity the day I came across it on the beach. I was drawn towards it, like a sculptor looking at a piece of marble and seeing there's a sculpture in there.

GILLIAN SCOTT
Entrepreneur

Eildon Street looks out on to the Warriston Playing Fields and the wonderful horse chestnut tree that stands centre stage feeding the local children their conkers.

We were lucky enough to move here three years ago and I overlook this old fellow who grounds me each morning. It tells us of the changing seasons through its solid boughs and transforming colours. The local schools and various clubs play against this backdrop. The composer Fryderyk Chopin stayed at Warriston Crescent when visiting Edinburgh in 1848 and is said to have composed a piece called 'Spring' based on this view and our magnificent horse chestnut.

Last summer it lost two large branches to a ferocious storm. This got me thinking. These branches were around two hundred years old and, to me, precious material. I wanted to use these branches to connect people to the value of our trees and to appreciate the process of making, understanding. As this horse chestnut is so well known throughout the local primary schools, I felt that the impact of this story would resonate with them. To create something from these broken branches would connect them further with the significance of making and the value of trees.

We wanted to share with the community the story of this tree as well as the understanding of wood and the slow process of making. The branches take a year to dry – horse chestnut is a creamy coloured wood and has the ability to absorb moisture – and have wonderful burrs, these tiny swirling knots that are tightly formed around trees. They create some of the most beautiful wood-turned pieces, the tree's unique signature.

JANE LEWIS
Urban wassailer

The whole tradition of wassailing is about singing to thank the trees for last year's crop. It's about chasing away any so-called evil spirits. There's a little bit of threatening in there as well, sometimes we're threatening the trees that if they don't produce well you'll chop them up. The songs are mostly quite light-hearted and fun, but the process of singing to the tree also establishes a relationship between us and a source of food that we often cut off from.

NOTHING CREATES LIKE A TREE

What we make from trees isn't always created from their material substance. Sometimes it's from their inspiration. Artist Victoria Crowe is known for her iconic images of trees in winter, bare-branched against snow or sky. When I asked what moved her about trees, she said, "It's an identification with another growing form. Throughout human history we've always had trees around us and there's been a lot of research done now on how trees communicate to each other at a level we never understood. Trees and nature form a kind of pantheism for me."

What Victoria saw in trees was "almost the creativity of the natural world". She added, "One of the trees I've painted, a contorted hazel, and seems like a convoluted intellectual argument. You can go that way, or the other way. I love it for that – it's very complex."

VICTORIA CROWE

Artist

When we moved to Kittleyknowe in 1970, the winters there were very snowy and stayed for a long time. It was a tiny little hamlet. All the trees were silhouetted against the snow and our little field that went down had this beautiful beech tree whose shape and movement described the prevailing wind. That was the first tree I drew from my studio window. Then I painted the twin beech trees. They were silhouetted against the snow and included our neighbour, Jenny Armstrong, who was the shepherd we used to see going about, as a sort of cipher really for scale. She was very much a scale indicator.

That little row of trees was almost like a demarcation line between the world of us living there and the world of the paintings and the landscape and that language that was building up for me. Then there was the real world of going to teach at Edinburgh college of art.

It was a very painful time when we left Kittleyknowe, because our neighbour and the man who owned the farm at the top died within days of each other, and developers moved in and so the whole spirit of the place was ruined. A lot of the land behind the beeches was dug up for peat extraction and coal mining. Everything was despoiled. I didn't go back to Kittleyknowe for a long time.

The focus of our lives changed, as all lives do. In 1994, my son, Ben, was diagnosed with cancer. He died in 1995, aged twenty-two. It came very suddenly really because he'd had a

sore mouth for a while and was going to see his dentist. I'd never heard of oral cancer, especially in young people – they don't normally get it. He never smoked – apart from the odd fag behind the bike shed and he was just a student drinker. My work as an artist continued, more aware of the immutable and enduring qualities of nature set against transitory and ephemeral human life.

I have only once painted a tree with leaves on. I did a series of paintings of trees and landscape at twilight. In them not only was I taking the leaves off the trees, but I was taking the light away from them. One painting was called "Landscape With Hidden Moon". On one level I was trying to see how dark you could go and still see the trees. But on another level it was very mysterious and quiet with all these huge trees and the silence and the light going down and these things still being there – very powerful feelings and associations that whatever is stripped away, the presence and force of the natural world is still there.

We have had a lot of difficult experiences. That's why it's hard for me when the painting isn't really flowing well. That's how I deal with those things – without it becoming too much a therapy. It exists as a kind of conversation with myself.

When I was diagnosed with cancer, I thought, I can do this. Ben went through this. There was the thought, too, that Ben went through it and didn't survive. Then there's part of you that says, Oh but it's not really that serious. I look back on it now and think, yes it was.

When I was going through treatment, the trees were still there as these visual reminders of this rich inner life that they had given me, even though I couldn't quite harness it. It's been a full year since then – an acknowledgement of ageing, of getting older, of being more fragile in some respects, but also stronger in others.

KEVIN DAGG
Artist

I am fascinated by the life cycle of trees. From collecting the seeds and potting them up, all the way through to them dying and rotting in the ground. As a sculptor, wood is my medium. I carve heads and figures but also use branches to represent lungs. I am just drawn to the material.

The process of decay has become my most recent journey and has drawn me towards the world of lichen and mushrooms. A world in miniature with its own battles and challenges fuelled by the energy of the tree, returning the tree to the earth and locking the carbon in the soil.

VRONI HOLZMANN
Artist

I've done a lot of work that relates to trees – photographing them, working with wood, piano compositions inspired by them. I came up with an idea for an exhibition which was to take photos of trees and then find the wood of the same kind of tree and make it into a frame. So I had a photo of maple bark surrounded by an actual maple wood frame or cherry branches surrounded by cherry wood. I took a lot of photographs not only of trees but also close-ups of the bark. I found all the different textures and colourings so interesting. I made those frames on my father's old workbench which I had brought from Germany to Scotland. My father was a cabinetmaker. I also trained as a cabinetmaker. We're even called Woodman.

Also, when I'm curled up in my bed, I rest in a space that my grandmother was resting in, an old antique bed brought from Germany, made by skilled craftsmen working with trees at the time, a hundred years ago and still solid. So these trees from a long time ago gave my grandmother sweet dreams and now they comfort me in my sleep.

FORGOTTEN GIFTS

Some of that knowledge of what trees can give us, and how to create from them, is in danger of being lost. Robert Penn, the author of *The Man who Made Things out of Trees*, told me how, he first learned the uses of an ash tree from a local farmer who told him how if he lost a wooden tool handle on his farm he would go to the pollarded ash and take a stem to make a handle. "The ash," Robert observes, "is the tree that we in the temperate parts of the world have been most familiar with over human history, absolutely integral to our progress of our species." Yet, he says, at this point in history "up to 90 per cent of people if you asked them, wouldn't be able to think of a use for ash beyond firewood".

Though we may have lost so much of this knowledge, we haven't entirely forgotten our debt to trees. Arguably, in these times of the pandemic, we are even more aware of it. There's no escaping the huge debt we owe trees as living carbon sinks removing our greenhouse gas emissions and creating the oxygen we breathe. We owe them, and we know it.

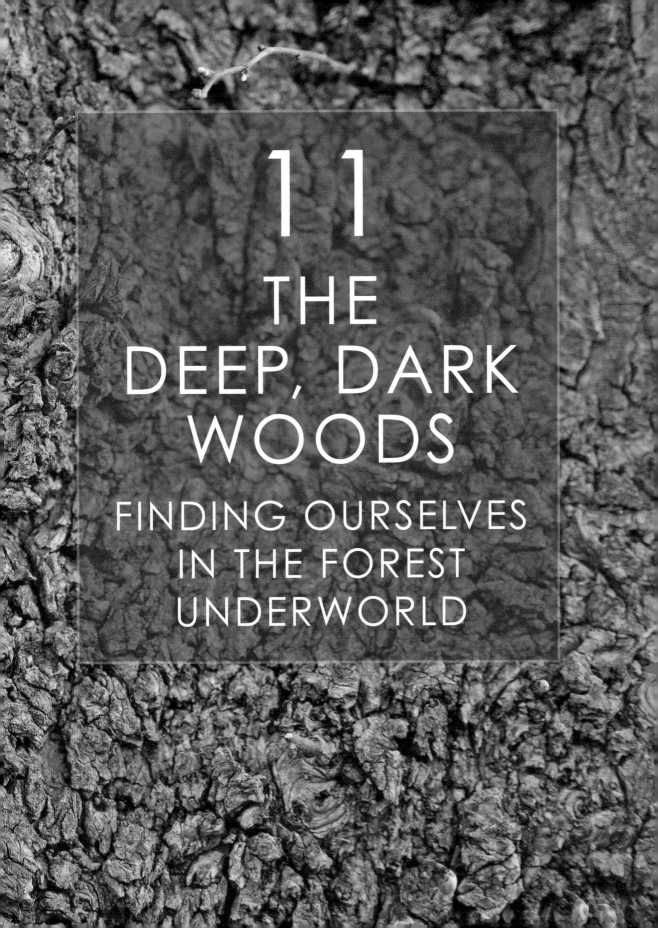

11
THE
DEEP, DARK
WOODS

FINDING OURSELVES
IN THE FOREST
UNDERWORLD

some forests remind you of other ones / existing in myth
& song / remember the thirteen-year-exile that the five brothers lived
& two women / hidden & lacquer-painted, switched & mimicked

PRATYUSHA, 'if still forest (winter)'

Let's not be too pretty about this. The forest has its dark places. These are one of the reasons we go there – to hide, to meet in secret, to stumble across, in the darkness, some element of ourselves we have hidden even from ourselves.

There are forests out there that let in scant light, whose floors are scattered with lumps of dead wood, pricked through by phallic fungi, and which do not let you escape without covering you in its sap and twigs and pine needles. These are woods that feel primordial. They call up thoughts of Tolkien's Ents, or of Grimm's fairy tales. Lost in their shade, we can feel as if we are more our animal selves, than our cultured selves.

For nature therapist Stefan Batorijs, the forest is "a place where meet our unconscious". He says, "When you enter the forest you're leaving one world behind and you're entering a new, different, magical world, where anything and everything is possible, where time doesn't exist, where everything is in a state of potentiality."

Plenty of those I talked with about their love of trees testified to the beneficial power of a walk through the deep, dark woods. Photographer Jannica Honey recalls how, in the wake of unsuccessful IVF, she started a project which involved taking photographs of women naked in the woods at twilight. "It is a place where you can encounter your own shadow side, or subconscious. For me, heading into the woods became a practical exercise in entering this space."

Many spoke to me of the woods as portals to other parts of ourselves, to altered states, to epiphanies. Some travelled there aided by drugs. The actor Tam Dean Burn spoke of being up in a tree at music festival. "I was tripping," he said, "just flying and sitting up a tree and able to recognise just then not simply that it was this big thing to sit on, but they are living beings that are going up towards the sky. The energy of trees is incredible. Once your head goes there, you realise how powerful they are."

JANNICA HONEY
Photographer

I started to take the photographs of women naked at twilight that would become my project "When The Blackbird Sings" in 2016. I had just gone through two cycles of IVF and it was awful. It is really tough to go through fertility treatment at the end of your fertility and I found the only thing I could do was to go out for walks. I realised when I was walking a lot that it was during the twilight that I found most comfort and peace.

During the daytime it's all bright and shiny and we're supposed to be productive and work, so twilight was the only time where I had that kind of peace and I could feel like the wheel of life was moving forwards. The seasons are changing, the water is flowing, the buds are becoming leaves and then the leaves drop off.

Creating these photographs came from the same place as trying to create a life, a child. The child grows in your womb, which is like that open space, a hollowness within women, which if it doesn't carry a child feels like it's just there, resting.

When, after my second IVF, I had a menstruation and knew I hadn't got pregnant, I went down with a friend to the Water of Leith to burn some intentions. It was the new moon and my friend and I did a shoot, with her naked, and when I came back I looked at the pictures and they gave me an immense comfort and I kind of decided *I can create something else.*

I headed into the woods because it was all about heading into the underworld, into the subconscious, the womb, the wild space where things move around like they always have done. It is a place where you can encounter your own shadow side, or subconscious. For me it became a practical and logical exercise in entering this space. I also found that during twilight, the light itself, or nature itself, would hold the whole session. It felt like wasn't up to me when I wanted to start the shoot or finish shooting. Just as it is with fertility, it was nature that estimated the time for me.

I would do the photoshoot in the blue hour, which is actually really a blue fifteen minutes. You're in the liminal space of twilight, where actually the air becomes filled with something. It is no longer clear. It's more than that; it's almost like the light itself touches your skin.

The word in Swedish is "grumligt" and that is when you have muddled waters. I find that light becomes muddled and in this muddledness, I think, if we believe in other worlds, the veil of what we see here in daylight is being pulled aside. I totally feel like the direct lines between the different time zones and the world are shifting a bit.

In the liminal space of the woods, I also found the women that I photographed would have conversations with me about abuse, rape, miscarriages, fertility, pain and sorrows, not because I opened up those conversations, but because nature created a space for it.

STEFAN BATORIJS
Nature therapist

Why isn't anybody talking about the truth? Why isn't anybody talking about anything other than what's nice with regards to trees and nature? If you go out into nature and sit there and just observe, what's presented to you is not about anything that's nice. It may appear nice in our anthropocentric "Disneyfied" perspective, but there is no sentiment in nature. I have just spent ages watching an ant carry a dead ant in its jaws over all sorts of obstacles.

Everything in nature is about a beautiful efficiency and opportunism and sex, so why do we focus just on these idyllic pictures of trees and sunsets? Why is nobody talking about sex and eroticism in nature? Why is nobody talking about the perceived shadow aspects of ourselves and our behaviour? Why is nobody talking about things like death, decay and defecation? Have you smelt how sweet the rotting leaves and humus layer of the forest is?

There's a lineage of people and writers who have an understanding of the more dangerous and lustful aspects of nature and the woods. The poet, John Clare, alludes to the eroticisms of nature, as does Nan Shepherd in her wonderful book, *The Living Mountain*, which is just full of latent sensuality.

I think people sometimes have an aversion or an inability borne out of separation for engaging with the darker side of nature. But if we are going to talk about the forest and the woods and the trees, we need to make sure we include that in it. We must not sanitise it.

Most of my explorations into these depths have been through exploring Jungian interpretation of fairy tales as being connected to archetypes, and most fairy tales seem to feature a forest at some point in the story. But also maybe more pertinently through the vessel of my body, my sensing apparatus, feeling the textures and elemental presences that feedback to me that I'm alive, on this ontological quest into the nature of being, and being of Nature.

The forest is a place where we meet our subconscious. That's what I teach my trainees. They don't necessarily need to express it explicitly with their clients but I ask that they develop this understanding. This is a really key aspect of my work, that when you enter the forest you're leaving one world behind and you're entering a new, different, magical world, where anything and everything is possible, where time doesn't exist, where everything is in a state of potentiality.

The way that we do that is that we step over a threshold. So when we're designing or choreographing our forest bathing walks, we look for that place that is the threshold, the entry into the magic world.

It can be a little bridge over a stream into the woods. It can be a stile or a gate. It can be one of those trees where you have two trunks coming out of one base and it forms like a v-cleft. You have to squeeze between the v-cleft to get into the magic world, and that for me ties in with some of the underworld myths. In the

Sumerian myth of Inanna/Ishtar, she defiantly insists on entering the underworld with all of her royal regalia, symbolising the ego. But as she descends through the seven gates, her sister, who rules the underworld, instructs the guard that she is to be stripped of her authority, until finally she is naked and powerless. This is how we can enter the forest if we want to learn more and be healed.

LAURA ANDREWS
Artist

Stuart Woods are often quiet as most people bypass them and head straight for the Roman villa nearby. Quite frankly, the woods, ancient and mostly a mix of wych elm, ash and old oaks, are more interesting. Stepping through the wooden gate into the wood is like stepping through a portal. Wych elm folklore has it that elm is a link to the underworld and it certainly feels like that, as if the veil between this world and another might be particularly thin.

I discovered these woods when I was separating from my husband and took lots of very long walks to "work it all out". I felt like the woods were there for me, as if there is some sort of communication and connection with a wild and wonderful unknown which made everything seem a bit better. The whole of the body and mind are engaged when walking in there, especially when there are slippery muddy paths and slopes to navigate. There are answers to be found there if you're open to them.

Donna Giffen

DONNA GIFFEN
Performer and fitness presenter

The forest is somewhere I go if I've got anything to sort out. I go into the forest, not to think, but to ground and return to myself. It has helped me through difficult moments in life.

I have a memory from when I was a kid and it's always what I seek out in a forest. It was at Beecraigs Country Park and it's a certain path where it's beautifully dark and mossy. The rows of trees are neat – and I'm not a neat person, I like mess –but there is this beautiful dark path, in which, as you're walking through and, whatever way you glance, you can't see anything beyond the trees. You can't see the end of it. It's about being in the depths of somewhere.

I lived in Los Angeles for three years, from when I was twenty-six, and everything there is dry and quite barren, and to get to anything green and wild you have to drive for a few hours. It's a lot of concrete and manmade grassy areas, some parks. Eventually, though, I just had to go to a forest. And we went about two hours outside of the city, to woods of beautiful tall silver trees and I was just in heaven. I thought, I actually need this in my life.

Sedona in Arizona is one of my favourite places to visit, the energy is incredible, particularly around the vortex points. Something happens when you drive into there from the city through the desert – it's like you can't hear yourself anymore. You're tapping your head, thinking, "Hello, anyone there?" But you're gone. It's very peaceful. The same thing happens to me in the forest here. It's almost like the true self comes through.

HUSH. HERE ARE SECRETS

But the forest is also more than a place we meet our shadow selves. It's where real shadows go to hide, where the homeless make camp, lovers venture to have sex, those who transgress society's norms enter to satisfy their desires. Luke Turner, in his compelling memoir of Epping Forest, in *Out of the Woods*, describes how his research uncovered how people had long found in the cover of its undergrowth a hiding place for "a multitude of lurid activities".

When Zakiya Mckenzie was writer in residence for Forestry England in the Forest of Dean, she came across the phrase "fern ticket". In that part of the country, she says, to have your first sexual entanglement in the forest was to get your "fern ticket".

"It's definitely," she describes, "a thing that's hush hush in the Forest of Dean, and that that goes back to the 1800s. They actually used to hand out cards, little fern tickets. Locally it's a kind of known unknown, that we pretend we don't know. But everybody does of course know about it."

You don't have to go to some deep, dark forest to encounter this. There are copses, woods and leafy corridors in cities that seem to provide portals to this wild, earthiness, this upspringing of the subconscious. We can even connect to such energy through a single tree. I'm thinking, here, of a sycamore I know well, whose sturdy trunk seems to literally grow sideways out of the wall above our local river, the Water of Leith.

Go there, and you can see this is a place of frequent visitors. Its contorted and muscular branches are tattooed by glyphs. There are swastikas here, gang signatures, markings.

It's also a tree that people have visited for decades, as a friend of mine has testified to. "That tree," he said, "was a sort of a metaphor for my childhood. There was a time that's all I thought about, climbing it, but then there was a time when I conquered it and stopped wanting to climb it. The tree matched how I was growing as a boy and man."

The tree was even where he lost his virginity. "I'll never forget it because I took my shoes off and my shoes went in the Water of Leith. The tree's right next to the water and they fell in and started to go down the river. I will be remembering that to my dying day."

The girl was in his sister's class at school, older than him. "You're so naïve at that age," he said. "I won't have been the first person to have sex there, or the last. That tree has seen action. Put it this way, it's not a tree in the Knoydart peninsula, it's a tree in Leith."

LUKE TURNER
Author

As part of a sound installation I worked on in Epping Forest in 2019, we did an ecological survey of a 30x20-metre stretch of forest – and as well as all the different species like hornbeam saplings, we found booze containers stretching from back to the 19th century to modern booze cans and vodka bottles, as well as used condoms and underwear and hypodermic needles. I'm guessing the latter from some heroin user coming there. I thought it was amazing that in this one small square of the forest we found all this evidence of human activity and misbehaviour.

With my book, *Out of the Woods*, I started off trying to write a social and natural history of Epping Forest. It interested me why people were so connected to this place – partly because my family were from the forest, so I've got this personal connection and partly because it was rumoured that we were descended from an aristocrat who lived there. I started going into the London Metropolitan Archives to research the forest history, but then it morphed into this more personal story. That happened partly because it was a time of huge turmoil and depression and a huge mess in terms of a relationship that had gone wrong and the shame and difficult feeling I had around my bisexuality.

Nature writing is such a purity genre. It's all about humans in opposition to nature. It's about surrendering to nature and having this position of awe, respect and worship towards nature. It ignores this murkiness that I think Epping Forest, because of its proximity to London, really gives the opportunity to explore. People have misbehaved there for time immemorial, from back in the times of the highwaymen to when it was saved for the people of London in the Epping Forest Act of 1878 and people just did not do what they were supposed to do. They were supposed to go there and do the nature cure – make themselves pure and clean and get rid of the stain of the city. But instead they brought the city to the forest with them. There was this brilliant human activity going on, this very fecund, naughty behaviour. That still goes on and that's not talked about so much.

CLEMENTINE EWOKOLO BURNLEY
Writer

Trees were useful for the times when I needed to be alone. Trees were excellent therapists after my first period when I worked out I had to spend my life confined to a body that sprung regular, public leaks and there was nothing I could do about it. Trees said very little. They smelled of small lives and infinite patience. Even young saplings were vast enough to contain all feeling. It had happened in church. My horrified mother had marched me down the aisle. The only creature I wanted to see that day was my evil-tempered dog, Mascot.

CATHERINE INGLIS
Science teacher

I got very lost, once, in some woods at Symond's Yat, when I was working for the RSPB back in 1992. It was during the summer on a warm, balmy evening that I decided to take a walk. I left the caravan I was living in and walked into the nearby woods.

At the start, there was a discernible path and I just walked along it for a while. It got quite dark, initially because of the density of the trees, but more so because it was actually getting dark. I had no watch, or phone, and had no sense of time. The darkness sort of crept up on me with my eyes adjusting gradually to the dimming of the light.

I decided to make my way back, but it didn't take long to realise that I was stumbling about with no sense of being on a path or even being able to see it. I was lost. By now it was pitch black. No one would hear me and I was too scared to shout because I felt it would make me more vulnerable and exacerbate how alone I was. Panic was starting to well up inside, but I forced my rational side to reason that no animal would harm me and I'd be very unlucky to meet a man.

But I was also crucially aware that the wood was bounded on one side by a sheer cliff edge which plummeted to the Wye river several hundred feet below and I had no idea how near I was to the edge. It was that that made me decide to stop and stand still and try to stifle my imagination going into overdrive at every crack of a twig and animal rustling sound. It was hard but once I'd made my decision, I literally lay down where I stood and waited for dawn and the light. I didn't sleep for a long time.

I got back to the caravan at about 6 a.m. the following morning in watery grey light.

Hamish Napier

HAMISH NAPIER
Musician

"The Tree of the Underworld" is one of the tracks on my album, *The Woods*, and it's about the elm. It tells the tragic tale from the 19th century, about Allan, the son of a poor widow, Christy Grant. They were one of the many families working in forestry who made their home deep in the woods of Strathspey. Allan had charge of the Loch Eanaich sluice gates. In her 1898 *Memoirs of a Highland Lady*, Elizabeth Grant of Rothiemurchus describes a stormy night when he did not come back: "When evening came on and no word of him, a party set out in search, and they found him at his post, asleep seemingly, a bit of bannock and the empty flask beside him. He had done his duty, opened the water gate, and then sat down to rest. The whisky and the storm told the remainder. He was quite dead."

When I am writing my music, I might, as I did with my album *The Woods*, think, right, I want to write music about an elm tree. You look at an elm tree and it looks like something out of Halloween. It's got all these gnarly branches, and all the bark is warty and twisted and sometimes a branch will just fall off an elm tree on to somebody. Elm trees have this mystery about them. They've been known as the tree of the underworld and people make coffins out of them. So suddenly you think it's going to be in a minor key. It's not going to be a fast and furious piece like you might do with an aspen – which if you look at it is so full of life, its leaves are shimmering. When you look at the elm, you think it's something lower, deeper in energy, slower, minor, and suddenly your canvas gets more focused. There are still an infinite number of possibilities about what you could write, but you've just narrowed it down a little bit.

ARRAN SHEPPARD
Environmental arts performer

I've always had a bit of an affinity with oak trees – the gnarliness of them. Some of the characters of the stories I tell in the forest come from the faces I see in those old trees.

We see things in the woods. Our anthropomorphising minds project onto the twisting, contorted trunks and branches. The shapes and forms of trees can seem very human to our eyes, sometimes menacing, sometimes comical, occasionally erotic. I have a liking for photographs of trees "eating" objects – wires, fences, discarded bicycles – growing around them such that they look as if they are chomping through. My eye seeks out those trees whose branches clinch like lovers.

Folklorist Dee Dee Chainey, the force behind the popular weekly social media event #folklorethursday, told me she had noticed that whenever the Thursday is dedicated to tree lore, it buzzes.

"People seem to have a particular affinity with trees, evidenced by the way we see faces and figures within them. Many pictures of tree sprites and forest nymphs come to mind, with their limbs entwined with – or even becoming – branches themselves, like in the illustrations of Arthur Rackham. Entering the woodlands is like entering a place where time, and the normal rules of life, are often suspended. It becomes 'fairy tale time', belonging to a world of magic, of dryads, of Baba Yaga."

Walk into the woods and we bring another

forest with us – the one that exists inside our heads and is inhabited by a panoply of fears and yearnings. Lust is prompted by the forest smells, by its strange shapes. Horror lurks there, planted partly by stories, and I don't mean just those Brothers Grimm tales, also the more recent fictions, horror films like *The Evil Dead* and *The Blair Witch Project*, novels like Will Dean's *Dark Pines*. And I mean the true stories too, because it's not like there aren't tragic and horrific tales of bodies found or dumped in the woods. These things do happen – and they haunt us.

Of course, at the same time we know this isn't where the real dangers are. These days, Death, like everyone else, has moved to the city. And a forest walk, in the UK, is about as safe as life gets.

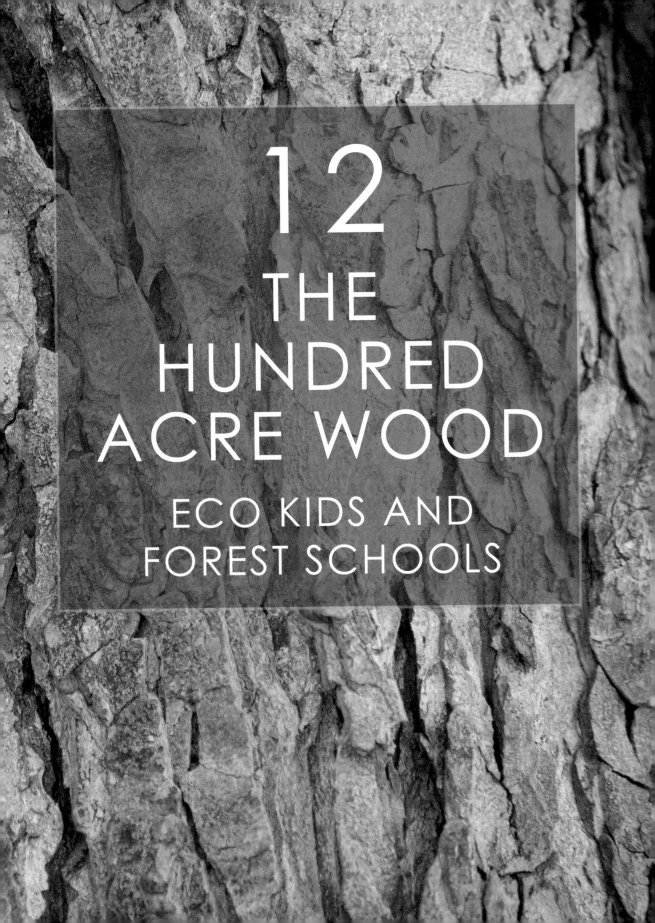

12

THE HUNDRED ACRE WOOD

ECO KIDS AND FOREST SCHOOLS

*If growing up means it would be beneath my dignity
to climb a tree, I'll never grow up, never grow up,
never grow up! Not me!*

J. M. BARRIE, *Peter Pan*

When we think of woods, or trees, often we think of children playing there – climbing high in their sprawling branches, gathering conkers from the ground, kicking through piles of fallen leaves. Woods are where the Gruffalo roams, where Pooh bear resides in his tree-trunk home, where Tolkien's Ents stride. Trees are home to fairy doors, they are where you can build a den or a treehouse high in the branches, where the wild things roar their terrible roars. They speak of the magic, mystery and fear of childhood.

WHERE THE WILD KIDS ARE

I always wanted my kids to be tree-climbers. I grew up in the countryside climbing trees myself and I wanted that for them too, just as I wanted paddling in streams, splashing in mud and making worm factories. I wanted it all the more when, while my youngest son was still a toddler, I read Richard Louv's *The Last Child in the Woods* – a book whose title always makes me think of these children as some sort of biodiversity loss. The author links absence of nature in the lives of children to a range of seemingly contemporary issues, from obesity to attention disorders and depression. It was Louv who coined the term "nature deficit disorder". As a result, when I met the organisers of a Glasgow playgroup, Nurture in Nature, inspired by them, and with the help of a friend, I set up an Edinburgh meetup group for families in nature.

In recent years, forest kindergarten, forest schools and outdoor learning projects have grown in popularity, alongside research around the benefits of nature contact. Among these projects is The Children's Wood at North Kelvinside in Glasgow, a small patch of woodland and meadow which was set for development, until the community, led by campaigner Emily Cutts, fought back – and won.

I visited the woods back in 2015, at a time when Cutts was fighting off the developers. At that time, the Wildlife Trusts launched its Every Child Wild Campaign. Its supporter, Sir David Attenborough, declared, "Contact with nature should not be the preserve of the privileged. It is critical to the personal development of our children." The Children's Wood seemed a place that was hitting entirely the right note.

Julia Donaldson supported the campaign, as did the actor, Tam Dean Burn, who was a local. Part of what he has loved about the woods, he says, is that his daughter "has been able to just run free".

One of the reasons, Cutts says, that she was so keen that the community kept the woods as local resource, was income inequality. The Children's Wood is at the boundary between an area of wealth and another of deprivation. On one side is Wyndford, which, on the index of multiple deprivation, scores poorly at only one point, and on the other is Kelvindale and Hyndland, which both score a prosperous ten.

On one of the very first children's forest school-type sessions, she says, one of the boys told her he had never played in an outdoor space before. "He was primary school age," Emily says. "And I couldn't believe he didn't play outside at all. He gave the classic answer that he just played his computer games with friends. To me that that was so motivating because he loved being out in the woods. We've all heard the theory that children aren't getting outside. And this confirmed it."

The feelings we develop around nature in childhood carry into adulthood – allow them to develop and they will be returned to later in life. That continuity was described to me by photographer Jannica Honey, who observed how being around Scots pines allows her to "plug into" childhood feelings and to her schoolyard experiences back in Sweden.

"I remember we had these huge two pines and around them, the kids would play, and during the winter a huge hill of snow collected around them. Then in the warm Swedish summer, the bark really heated up, and I remember the feeling of the bark underneath my hand. I remember the smell. Even if some of the things in my upbringing possibly were quite dysfunctional and it wasn't all happy, my memories of having connected with nature have an underlying tone of contentment. It's almost like I plug into something – a connection to something much bigger than just a tree."

ROLL UP, ROLL UP FOR THE NATURE CIRCUS

Research, meanwhile, is showing us that the woods are also good for the well-being of children. "A large-scale Canadian study looked at the potential psychological benefits of nature-connectedness in children," psychiatrist Charlotte Marriott told me, "The study showed that engagement in outdoor play and an appreciation of the importance of feeling connected to nature were associated with decreased prevalence of psychological symptoms: feeling low or depressed; irritability or bad temper; feeling nervous; and difficulties in getting to sleep. This suggests a potential protective effect of spending time in nature against symptoms of poor mental health."

Bringing children and adults into the woods is also one of the things that environmental activist and trapeze artist Lucy Power is attempting to do. With her partner Arran Sheppard, she runs Rowanbank Environmental Arts & Education, which delivers forest schools, teacher training programmes and circus events, principally in urban woodlands and parks. They work in some of the most deprived areas of Scotland and utilise even the smallest cluster

of trees. "When we run Forest Schools and perform our Forest Circus, the woods become this magical space."

As well as her Forest School training, Power also undertook training in a Swedish approach to outdoor learning called Skogsmulle. The process sees nature connection as a flight of steps which starts with children simply enjoying nature, moves through observing it, then caring for it, and ultimately has children perhaps involved in campaigning to save nature. "Often you expect children to be on the third step already, to care for nature, but they've never had the opportunity to properly experience those first two steps – that time and space enjoying nature and observing it. More and more children are disconnected from nature and don't have that crucial time

in nature at home or at school to enjoy and observe. We realised that's what we've got to give. We enable that."

More and more, people talk about children's disconnection from nature – that they climb trees less, roam over smaller areas. But that's not all I see. For I'm also aware of a new generation of young people who have found a whole different level of connection, powerful and intimate, but also coloured by the knowledge of the climate crisis. They might not be the majority, but they are there, and very vocal. Some, like Dara McAnulty, author of *Diary of a Young Naturalist*, write books rich with knowledge. Some of them are showing us how to be with the trees. More interesting than what parents have to say about kids and trees is what the young people have to say themselves.

POPPY AND LILY
13 and 15

We have a tree we call the Spider tree that we have climbed from as soon as we were able to at about six years old and we've now moved on to the big tree up the back of the garden. During lock down we climb it to hang out and chat. There are six spaces to sit. You can see the whole of Edinburgh from it.

We like to climb trees because they support life. Climbing them gives us freedom. They grow with us and become familiar.

Climbing is a great release and challenge, and it's always calm at the top. If I'm angry or frustrated climbing and sitting in the big tree calms me.

ELIAN
11

It all started because I wanted to live in a forest. I can't remember where it came from, but I kept talking about it. I talked to my friend about living with him in a forest when I grew up and I started getting plants in my room to make it feel more like a forest. There's even one that is like a vine which goes all around the ceiling of my room. I now have twelve plants all together so it feels green and like a forest.

Then I got interested in tree houses and watched programmes about building amazing tree houses. My mum and dad built one in the garden but it was on stilts because we didn't have a tree it could fix to and it was quite small. So I kept talking about a real tree house. That was when Mum and Dad decided to buy the wood because they thought it would be fun and good for us as a family to be able to go to the same place to play in and take care of nature and be able to build things.

I like being in a tree. I feel like a bit like the woodland elves in *The Hobbit*. The trees that are good for climbing are the ones that have lots of branches but if the trunks are thin and they don't have many branches, they can be good for shinning up. I like to jump from quite high up in a tree. I also like to hang and swing from the branches.

I don't have one favourite tree but a lot of favourite trees.

For example, there is a curly willow in our garden which we planted when we moved here and it's now big enough for me to climb. When it's windy, being up in it is like riding a bucking bronco.

Last year, we made a tree house that is also in the pine trees. It's in the part of the wood where I really wanted it to be and quite high up. It has a pine tree growing through the middle. It has a rope ladder to go down and up. It has a balcony with a great view of fields and the valley where you can eat your lunch and look at the birds. We made some nice bean bags to put in it and took lots of cushions and some home-made decorations. We slept in it in the autumn. It was so cold that both my Mum and I dreamed it was snowing.

NOGA LEVY-RAPOPORT
18, climate strike leader

In our communal garden there's a huge tree that we have a swing on. I was always a massive tree-climber. Anywhere we went, any tree we saw, I would hop on it. I remember just obsessing about this tree because it was so tall, it's still massive. One year, when I was about eleven or twelve, I kept trying to get higher and higher and test the limits of these very spindly branches, and I always recall this moment of getting to the top and just having no fear at all.

It's enormous and I've no idea how I was not terrified, but I just did not feel afraid at all. That is something really special and you don't really get that anywhere else. Up in a tree there's that feeling of the wind in your face and looking out over the world. There was really a sense that I could now see the city and look out over my garden, out over this great expanse of the city. That's a really special feeling.

I live right in the centre of London and it's increasingly hard to access green space. I was never one of those kids that was obsessed with species of flowers or birds. But actually just that real sense of freedom and expansiveness, that nature has. It has always made me feel quite enfranchised in those spaces. It's given me the confidence to exist in those spaces, in mirroring the way that nature just calmly trots along. It's such a gift to be able to see out of my window and that has kept me going.

I used to be shy and struggled with my mental health. When I first heard about these climate marches and saw how driven it was by young people, I thought I'd never seen anything like this before. I'd never been in a space where young people had said we're really angry and we're going to transform that into action.

When I turned up to the first strike I remember thinking, this isn't enough. We have to do more than stand on the grass and chant. I turned to a kid that was next to me, who had a megaphone, and I said, "Hey, I don't know you, but can I borrow your megaphone?" I grabbed it and said, "Follow me. We're gonna march down this road." By the end of that day there were five thousand kids following me on to that road.

MYA-ROSE CRAIG
18, birder

GEORGIA
6

The woods where I live are havens for wildlife, and during quarantine we've taken the opportunity to explore even more. One thing we have really been enjoying is a pair of ravens. They have been at the edge of the woods, really near our house, and they're amazing. We've been checking up on the nest and like seeing them fly around, scaring off all the predators.

I set up an organisation called Black to Nature. One of the things we do is run nature camps. Many really positive benefits for the kids come on these camps. It's really important for mental and physical health that we have that connection and engagement with nature. But also I think that connection is more important now than ever because of the ecological crises we have at the moment. We can't expect everyone to be on board when a percentage of people haven't even ever experienced that connection to nature in the first place.

I like blossom trees because they are really pretty and my favourite colour.

CAYDEN
8

I often climb the trees in the local community garden, Leith Croft. I like them because they are things that live. I'm less scared climbing a tree which is really high than I would be climbing a climbing wall with a harness attached. Trees feel safer because firstly if you fall there are loads of branches you could catch, but also because they are different – they are alive.

ELSIE
8

ROSELLA
6

I like playing in the woods because I'm surrounded by nice flowers and trees and they make me happy. I love watching squirrels and animals running about. My favourite is spotting ladybirds.

I like climbing trees because when I get to the top I like being able to see how far I've climbed up. Also I like being able to see the other heads of the trees and the people below me – and I like touching the branches.

HOLLY GILLIBRAND
15, climate activist, @HollyWildChild

Several years ago when I was visiting family in New Zealand, I came across the biggest tree I have ever seen in my life. It towered far above the surrounding coastal landscape and, being the energetic young girl that I was, I climbed it. Many of the branches were the width of a tarmac pavement and its lower limbs stretched towards the sand for support. It must have been hundreds upon hundreds of years old; weathered, battered but unbeaten. This tree was a pōhutukawa tree.

The pōhutukawa is the New Zealand national tree and an important symbol for the indigenous Māori people. In Māori mythology, the blood of a young warrior who died while attempting to avenge his father's death is said to be represented by the bright red flowers of the tree. My encounter with this particular pōhutukawa has stayed with me ever since. It taught me how indomitable the natural world can be.

But despite all efforts from environmentalists and activists, nature is in trouble. We are in the midst of the sixth mass extinction with up to two hundred species going extinct every day. If we want to preserve the life support systems that make up the foundation of our civilisation, we can't ignore the decimation of the biosphere any longer.

Last year I decided that every single Friday I would strike from school to bring attention to the climate and ecological emergency and it was my connection with the natural world that drove me to start my activism. You cannot stand beneath a magnificent natural wonder like the ancient pōhutukawa without feeling awed at the immensity and complexity of life on Earth.

ELIAN
11

HOW TO CLIMB TREES

1. When you climb trees you should always make sure you can get down again. You don't want to be at the top of a tree with no way to go down.
2. Before you climb always make sure the tree is stable and not rotten. You don't want a tree to fall down while you're halfway up it. You should check the branches you use are strong enough to hold your weight as well.
3. Shinning is a good skill to have while climbing trees. You put one leg on either side of the tree and one arm on either side, then push up with your feet and use your hands as supports.
4. Efficient clothing for climbing trees is tight clothing that doesn't snag on a branch or a twig. Don't wear your favourite clothes either because you don't want them getting ripped.
5. If you ever get hurt up a tree, don't keep going up – just come down safely.
6. Pine trees are fun to climb but sometimes there aren't enough branches.
7. Birch trees are normally thin so they are good to shin up.
8. A handy tool for climbing trees is the branch hook. It's useful for when you can't reach a branch. You can make a branch hook by using a strong piece of wood in a hook shape.

JUST SUPPOSING

The words and actions of young people tell us a great deal about our evolving relationship with trees and the natural environment. Though surveys suggest that statistically children may be less likely to know an oak from a beech, there are many young people out there who are highly educated in the workings of the living, and non-human, world.

For many of them the risk they think of when pondering trees is not that of falling from them, but of losing them, not planting enough and of the catastrophic events predicted if our temperatures keep rising. Because of this knowledge, their feeling around nature, as one climate-striker told me, can be "bittersweet". But it's also what makes them fiercely passionate defenders of the natural world.

I'm reminded of an A. A. Milne story in which Piglet asks, "what if" a tree were suddenly to fall on top of himself and Pooh? His bear friend's answer is, "What if it didn't?" I can't help but think of their imaginary scenario as a metaphor for the climate crisis. Piglet is asking what if the climate warms disastrously. And many of us adults, like Pooh, have got by with saying, "What if it didn't?" But we can now no longer get away with simply saying that. In fact, children – and, to be fair, many adults too – are not saying, "What if?" Rather, they are saying, "The tree is already falling, we need to do something now."

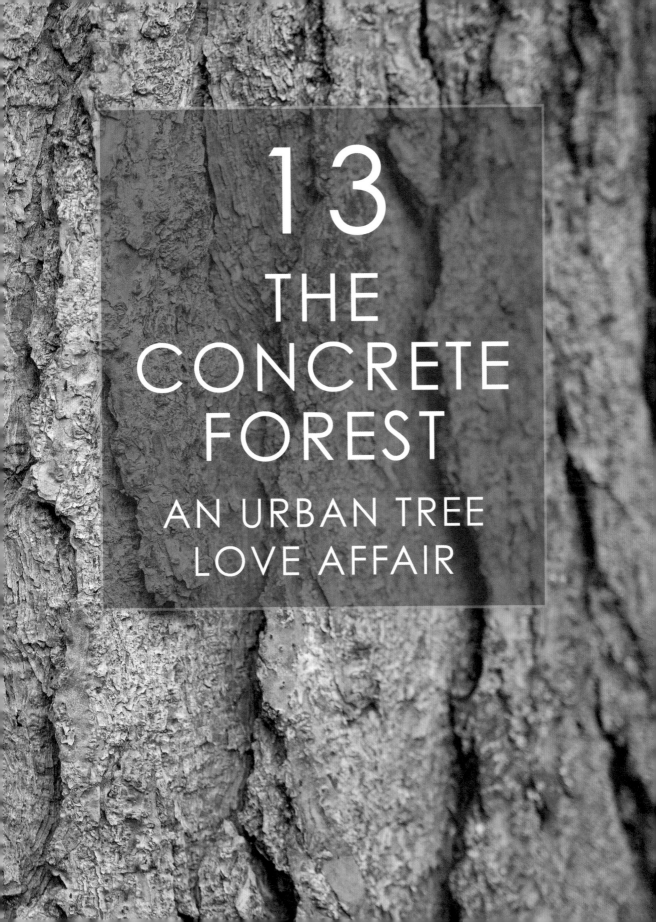

13

THE CONCRETE FOREST

AN URBAN TREE LOVE AFFAIR

The trees along this city street,
Save for the traffic and the trains,
Would make a sound as thin and sweet
As trees in country lanes.

EDNA ST. VINCENT MILLAY, 'City Trees'

The urban life is marked by trees. Street trees. Park trees. Trees spilling over fences from people's gardens. Trees sprouting through concrete. Our relationship with these trees is very different from a tree in the rural landscape. When we look at an urban tree, we know that either someone has planted it, or it is there against the odds, one of those outcrops of nature that has pushed through the cracks of our human carapace. We see them as we go about our daily business and rush. Sometimes we become attached without really quite registering their presence.

When I first moved to the area I live in, Leith, with a small baby in tow, there was a walk that I used to do. It would take me out to the bustle of the local Kirkgate shopping centre, under the shadow of towering high rise, a brutalist listed building but also in the second worst category on the Scottish Index of Multiple Deprivation. I knew why, in spite of the concrete, I loved that walk. It was because of the trees – the avenue of lush rowan, maple and cherry. The trees brought the place a kind of glory.

Research is now acknowledging and proving that trees enhance our health and everyday wellbeing. The simple fact of living in a city has been found to impact our mental health. As Charlotte Marriott describes, "Many studies have shown that urban dwellers react differently to stress from their rural counterparts and have a far higher risk of developing depression, anxiety and schizophrenia, although this is of course multi-factorial."

Increasingly studies reveal that green space and tree presence are not just a question of aesthetics, but also a public health issue. The presence of street trees has been found to decrease the risk of negative mental health outcomes. Research by Professor Richard Mitchell at the University of Glasgow has shown that access to green space also reduces the rich-poor divide in health and mental wellbeing outcomes. Socioeconomic inequality in mental wellbeing, he found, was 40 per cent narrower among people reporting good access to green or recreational areas compared to those with poor access. We cannot underestimate their value of trees has a presence in our day-to-day lives.

Often communities know that. Some living on one of the streets I would walk down with my baby told me how trees had helped them through lockdown, and also how, when the cherry tree on the corner was suddenly removed by the council, they were distraught – so much so that they planted a small cherry sapling in its place.

CAROLE WRIGHT
Community garden manager

My favourite tree from the orchards I've planted is a fig tree on the Peabody estate I live on in zone 1, London. We got funding to add an orchard to the community estate here. We asked for apple trees, nectarine and a plum, and then when Peabody's contractors, Gingko, turned up, they said, "Oh, we've got this fig tree that's been lying around in our yard for three

years – so we got a fig tree a well." It's got pride of place in the garden and it has come into its own over the past few years – there's been over one hundred fruit on that tree. People love them. It's all I can do to stop one particular family that live overlooking the garden, getting in there and enjoying them all.

It's the third orchard I've planted in SE1. The others have been on two housing estates. One is on a housing estate, off Blackfriars Road, and that is a Tate Modern Commission called Triangle Garden, which is thirteen years old this year.

This Peabody estate garden is a gem. Every plot is absolutely different. It's a little sanctuary, particularly at this lockdown time when our playground and our basketball court are all off limits. People can walk round the square. And the striking feature here is that we've got London plane trees from the 1870s. They were pollarded recently so they give these eerie shadows when the sun hits them. It's staggering. People come in and they're like, "Oh my gosh, the height of those London plane trees!"

I started community gardening at a difficult time in my life. I was coming from a women's hostel and I had two and a half of years of deeply unpleasant circumstances. I came across a garden and I was shocked because it was where my grandparents had lived and I hadn't been there when they were there. I literally walked round with my mouth open. The had eight raised beds, fruit trees along the railings.

I said, "What's this about?" The answer was, "Community garden. Would you like to come? You can share a raised bed." So that was my introduction to community gardening right there and then.

That was such a tough time, but my difficulty ended the minute I got the keys to this flat. The trees were the first thing I noticed on this estate, because you leave Blackfriars Road and you can just see the trees above the buildings, which are stunning, and then you walk through to the courtyard, and you're like, "Oh my gosh."

I started volunteering as a gardener and that was the turnaround. My first community gardening job was Brookwood which was a baptism of fire.

They said they wanted it done in six weeks. I said, "Where on my CV does it say – can create a garden in six weeks?" But what I do know is how to deal with people having grown up on a housing estate and I know you have to treat people right to get them to treat the plants right. That's always been the way. I let people know, "Please look after this space. Be mindful of how we treat each other in that space. Because it's precious."

I think that was a turning point. That made me know the power of greenery, of nature, of looking at things grow, of looking at people grow – just knowing that through gardening I could deal with children, with young people with ADHD, people with depression, people with substance abuse issues, people who have been threatening.

Landscape plays an important part in a city. When I walk about and see beautiful gardens I don't see why we can't have those kind of features on a housing estate. I grew up going to parks. I grew up running up and down

outside, particularly in the school holidays, in Kennington and Brixton.

We climbed the park trees – park keepers weren't amused. We would also climb on top of walls. I don't think I spent a day at my school in Kennington not climbing on a wall. I think my brother and myself were the only kids in our group who didn't break something and end up in St Thomas's hospital.

My family were into gardening. My grandfather, Eric Wright, who passed away about a decade ago, used to manage an estate in Jamaica. He was the foreman on an old colonial estate – he managed land and people and a bakery. My grandparents lived on the Wyndham estate and then the Elvington estate in a tower block. They had lots of plants in their homes, even to the last, the last place my grandfather lived, they had spider plants, really 1970s plants, but really green, and he had allotments. It wasn't until my grandfather passed that I knew, how good he was in his allotment. When we were sorting through his belongings we found he had won allotment keeper of the year. Never knew.

There are many other trees I love. There's a London plane tree in the Brookwood community garden. I remember when I first went to Brookwood, a resident ran out of a block of flats opposite because she thought we were looking at the London plane tree. She said, "I hope you're not chopping that tree down. I'll get up a petition." She properly came out and heckled us. I said, "Actually we're not going to bother it, we're looking to do a community garden." And she was one of the biggest supporters of the project from that day – once she established we weren't going to chop down any of the trees. People loved those trees on that estate.

CAROL MURDOCH
Outdoor education consultant

I grew up in Muirhouse, north Edinburgh, in the 1980s and 1990s when it was a bit of a no-go land, before the new developments down there and when it really was akin to *Trainspotting*. That was back in the days where we really did not realise the difference greenspace made to an area and to mental health. There was very little green space. I only really remember there being four trees around where I grew up, three of which were in my school playground.

I went to Silverknowes primary school. We had a raised coffin area, which was a humongous raised garden, around waist-height, and which had a tree. I seem to remember it being a sycamore or oak. Another tree stood not too far from this one, on ground level and it was also a sycamore or an oak. They were old trees, older than the school. The other tree in school was a blossom. Every year it would be beautiful. I still remember being a child running around the fairies with that blossom.

When I turned six or seven, we moved to a main-door house in Muirhouse Park. These were rare. Most of the houses in the area were flats. But this had the most beautiful lilac tree in the front garden. We popped a piece of wood up there to turn it into our treehouse. Hours were spent in and around this tree. As a moody teenager, I would stare out my bedroom window at this tree. In an area with such horrors, murder, mutilations and drugs, these trees really were a hope. A hint of the natural beauty of the world. A hint that no matter what, beauty could be found, if you just waited.

René Sommer Lindsay

OUR CITIES NEED TREES

There are so many reasons to appreciate the presence of trees in cities. René Sommer Lindsay believes that rather than design trees into cities, we should design cities around trees.

RENÉ SOMMER LINDSAY
Urban designer

Our cities are increasingly dependent on a range of invisible networks of infrastructure that run through, or under, our streets and buildings. Wires, pipes and sewers that lie unnoticed, but take up growing space for trees. We need electricity, IT, heating, water, and so on to run our cities, but for every new layer of infrastructure we add, we are running out of space for nature in our cities. If we want to have high-quality urban nature outside of our central parks it has to be part of the mindset from the beginning.

Architecture firm Tredje Natur developed the Masterplan for Klimakvarter (the Copenhagen Climate Resilient neighbourhood project) and they had one idea that struck a chord and propelled the project forward. Instead of designing our cities around technical infrastructure and allowing nature into the gaps, what if we designed our cities based on nature and then let infrastructure squeeze in?

Instead of laying out streets, sewers and services and fitting in tree pits wherever there is any space left, we can consider the ecosystem services we want to ensure and then carve out space for urban nature that is deemed untouchable. Then roads, pipes and wires must find other ways to get where they need to go. It's not a situation we have now and it will take a while before we get there, but especially for brownfield sites I believe there is great value in being clear on what biodiversity effect we want to achieve and quantifying what that means in terms of acreage for green infrastructure and then sizing the roads and technical infrastructure to achieve that situation.

For cities like Edinburgh and Copenhagen, on which I have worked, the main climate challenges we will face are increases in the severity and occurrence of extreme rains and an increase in average and extreme temperatures. Trees are a perfect response. The right tree in the right location can soak up and evaporate large amounts of rainwater and free up sewer-system capacity. Well-established trees are a resilient form of planting able to withstand flash floods and droughts. A large urban tree can create its own local microclimate: offering shelter from the rain, shade on a hot day, and the simple nature of the tree means that in any weather it becomes a natural gathering place.

ANNA LIU
Architect

My appreciation of trees is something that has actually developed with age. I live in London, was born in Taiwan and grew up in the States. My childhood in Taiwan was very urban and my teenage years in the United States very suburban. When I went to college and found my path, which was architecture, that was in a quaint, liberal New England town in Massachusetts, closer to nature. Then I discovered my voice and identity in New York, one of the most intensely urban cities in the world.

The connection with nature developed more once I moved to London, after the birth of our son. We were going back to Taiwan and Salford quite a lot, to see my parents and Mike's parents. Rediscovering and enjoying nature

– the strong and abundant tropical nature of Taiwan, and the delicate and diverse nature of the UK.

Around the same time our son was born, we were teaching at the AA School of Architecture, and we were exploring with the students patterns of nature and human nature. Nature and human nature, often seen as opposites, as contradictory, are very much connected.

Actually, if you look deep into human nature, it grows out of nature. In the design unit we taught for four years, we looked at patterns of growth, starting with seeds in the first year to trees in the final year. We played with form and patterns through drawings and models, and learned about nature's patterns' inherent logic and structural strength. In parallel to exploring beautiful patterns found in nature, such as the Fibonacci series, we also explored how human activities create patterns through space. Not only do trees mark important historic points in the landscape, create oxygen and coolth, they are marvels of active systems of exchange – air, nutrients, microbes. The intimate relationship between trees and architecture can be established in several different levels: direct, biomimetic, psychological, symbolic.

TILOUZE SENNI
Carer

I like fruits and cherries are my favourite, but we didn't have cherries too much where I grew up in the east, south of Algeria, because the climate is semi-arid – hot, dusty and dry in summer and cold and dry in winter. But spring is the best season ever. My village was situated one hundred kilometres from the desert gate.

Cherries were very expensive there. So, in the community garden, the Leith Croft in Edinburgh, I planted a cherry tree. I would have loved to plant a eucalyptus, but it wouldn't have worked on my allotment.

When I was growing up, we had giant eucalyptus trees very close to each other. I used to play there with my brothers and climb in them. They were important trees, home to birds and insects. Our cattle and sheep went under these trees for the shade and we shaded under them. When it is hot the eucalyptus produce beautiful smell. In the UK, you won't understand the true feeling of a tree shade. It needs to be really dry and hot to feel the importance of a tree.

Planting this cherry tree in Scotland was about putting down roots for me. The first time I did something like that, it wasn't a tree – it was a plant. It was my first trip where I had seen the sea in Algeria, when I was about twenty. It was to Bejaia, and it was amazing. I saw a plant that has very big bulbs, a bit like an onion, but giant. I stopped there and we shaded under the trees and had our lunch break and I saw those bulbs and tried to pull them out, because I wanted one. I planted them near my rooms at the university. After three years I went back there and those plants were still there, When I saw that they were still there, I felt like, *Me, I'm still there.*

I believe in God and I have a strong connection with plants. I can't go to a place and see a plant that is not watered and not water it. I feel sorry for it. I'm working as a carer, and when I go to people's houses, I care for the person, but I find myself looking at a plant, wondering how long it has been since it was watered. When the water goes in there, I feel like the plant is giving me a blessing. I'm Muslim, and I believe I will get a reward after. I talk to the plant. I tell her, don't give it to me back in this life – give it to me in the life after where I really need it.

LUCY POWER
Trapeze artist and Forest School leader

I grew up in the middle of London. But I was very lucky in that there was an apple tree in my garden that I spent a lot of my childhood in. I had my first plastic, little orange trapeze hanging from it and I built a treehouse in it. I wanted to be a trapeze artist when I was a kid, but I don't think I really knew what one was!

My work with Forest Schools and environmental circus is very much based in urban woodlands. We work in extremely deprived areas, where the woodlands and parks are often abused and experience antisocial behaviour, and therefore are often feared and underused by families. What we're trying to do is turn that perception around. We hope that by bringing people there to experience nature's magic, they can begin to see it in a different light. It is a bit magic wand-like – when you start using a space positively it often stops getting used negatively.

EMILY INGLIS
Artist

We (my collaborator Rachel Owens and I) came across a bountiful apple fruit tree while exploring the landscape beneath the Gravelly Hill Interchange AKA the Spaghetti Junction. Throughout the changing seasons of 2017 and 2018 we investigated Birmingham's Spaghetti Junction, focusing on the nature that had grown up within its boundaries. Finding such a tree full of ripe red fruit in the centre of such a place felt mythical. We imagined it beginning as a discarded apple core thrown from a car window on one of the flyovers above sometime after 1972 when Spaghetti Junction was built. Up a bank we found the tree, unseen, off the path, out in the light, surrounded by motorways.

OUR LEAFY TRIBE

We love our urban trees. This love is why so many will fight for them. It's why the people of Sheffield would not let theirs go. The trees in our cities are key elements in people's lives – meeting points, pollution-absorbers, oxygenators, distractions from our hectic lives; they are friends, community elders, companions. We may not always notice them, but we will resist and put up a fight when they are under threat, and, when they are gone, we miss them intensely.

14

A WONDROUS WILD WEB

THE ROMANCE OF CONNECTION

*When we try to pick out anything by itself, we find it
hitched to everything else in the universe.*

JOHN MUIR

We now know, thanks to scientists like Suzanne Simard and popularisers like Peter Wohlleben, that trees in woods are connected by an astounding system of mycorrhizal fungi, through which they communicate, feed and support each other. It's a fact that many of the people I spoke to for this book would marvel over at some point in our conversations.

There are many reasons we are drawn to this idea of the so-called "woodwide web", but one stands out for me. There's something in the image of the wood that is almost as much fungi as tree, whose web underground is an entirely different organism. The image of these two forms of life from entirely different phyla and families, interconnected, reminds us of our own connectedness. If the tree, the fungus, and the whole forest are almost one, what are we really almost one with?

It reminds us that, far more important than our internet, is the fact that we are part of a biological network, which is one we often forget to think about. The tree outside our window is connected to us through the air we breathe. In the sunshine hours, it takes in our carbon dioxide and releases the oxygen back to us. The chemicals it releases enter our airways.

What we have in recent years learned about trees is turning our ideas about what matters on its head. If trees can be said to talk, as some interpret their communication, what does it mean to talk? Anthropologists like Eduardo Kohn are even telling us that communication and signs, mostly chemical, are the signature of all life.

Often when we talk about trees we tend to fixate on two issues. The first is how many we need, how we must plant trillions more to save the planet or, at least, ourselves. Then, there is the lifestyle fix; the fact that being among trees, which some of us call forest bathing, is increasingly recognised as good for us.

But if you take on board the fact that we humans are not separate, but are part of a world wild web of organisms, of complex interacting ecosystems, in which trees are key, these two issues are clearly not different. They are integral to the one principle of connectedness. In our urban, and sometimes in our rural worlds we can forget that.

PAVITHRA ATUL SARMA
Environmental researcher and entrepreneur

My grandfather had a very close relationship with trees and plants. He would talk to them. He would tell them his troubles. He would sing to them. Right from when I was a kid, I remember watching him and I would say, "Why are you doing that, Grandad?" And he said, "Trees talk to each other. They can understand and absorb your thoughts. So if you're nice to them, they will be nice to you. If you're nice to nature, she will be nice to you. But if you take advantage of nature then you are in big trouble." So that was drummed into me from a very young age.

But I think what happened as I grew older, because I grew up in a city of twelve million people, was that I lost the connection for a while, because it didn't seem very important. But I would always talk to plants. I would always feel like I could go and sing to them, share my woes and they understood. And the same with animals.

This was the aspect of spirituality and connection with nature that I had lost contact with until I started working in the forests for my first Master's degree in ecology. It was when I was doing research in the Western Ghats that I became aware of my connection with nature.

I designed a study to understand the relationship between tribal and non-tribal communities and their dependence on the forests for sustenance and their nutritional needs. The study took into account the type of forest areas and the impact of laws, policies and acts on food access. I was also living with the communities.

What really made me want to look into these dynamics was how amazingly interconnected the flora and fauna, and the relationships, synergy between people and the environment were. That's the fundamental definition of ecology and seeing it in action blew me away.

There was one young forest guide who would accompany me and we would go across forest areas in the Ghats where we would see the most amazing species of birds and trees, and as we walked he would tell me things like, this particular species is good for a woman's uterus. I said, "Can I write this down?" He said, "Yeah, go for it, write it down, but don't ever be one of those people who publishes their research and forgets about us."

In the end I decided not to publish my research as my guide's words sunk deep. I was aware of the politics within academia and the avaricious mindsets of senior faculty wanting to usurp and misrepresent research that wasn't their own I thought about the numerous discussions I had with tribal members about who benefitted from my research. There was so much knowledge that they all gave me. I remember my guide said, "Here the trees talk to each other, I don't know if you all study that. I'm not talking about the scientific aspect of it. If you stand very quietly in the middle of a glade you can forget everything and you'll hear a whisper, you'll hear and you can feel it. There are some things that you can't learn from books."

When I came to Scotland, I loved coming to this new flora. There's a rowan tree at our home. When it was sold to us, it had had its arms chopped off. When I first saw her, I couldn't understand why anyone would chop her arms off. She looked really bare. She looked like she was crying. I said, "This is a very powerful tree and she should not have been treated like that. We've got to figure out a way to keep her happy."

Every time we leave home and we are travelling. I will always say, "Please protect our home and stay safe." Sometimes the kids hear me talking, and they say, "You are nuts, Mum." I say, "You know how you are with your animals, well I'm like that with certain trees. When you talk to plants and animals, especially plants and trees, or sing to them, they do grow better." There is evidence around this and how trees "talk", especially with more research coming out on the symbiotic relationship between trees and below-ground fungi. But the problem is we're very dependent upon evidence from science and there's so much we don't know. This is why traditional knowledge is really important to preserve and understand.

SIR GEOFF PALMER
Scientist and human rights activist

Without trees, there would be no humanity. What I mean by that is that without trees, which produce oxygen, we would be dead. If we had no trees whatsoever, if we took away all the plants and the whole concept of photosynthesis, then human life would end.

I did an Honours degree in Botany and my research has been on plants. Plants have been crucial in my life. By looking at a plant, you can tell a lot about it. You can tell how it's working. I look at the trees on the hills where I live and I know that the tree in front of me is avoiding the wind – because of its position and because, if it didn't do that, it would break. Its physiology and structure are related to function. That's what I used to tell my students about trees. Understanding physiology and structure is how I made my breakthrough with barley abrasion used for making malt for brewing. Plants are a very good example of the relationship in nature between structure and function.

When I was a child, growing up in Kingston, Jamaica, the main tree I noticed, which was part of my life and the only tree I knew well, was the breadfruit tree in my yard. God knows how it got there. When my mother went to work in the UK in 1951, she left me living with her sisters, in a house with this enormous breadfruit tree, and it bore fruit, and we ate the fruit, and just over the fence, there was also an ackee tree which was mostly in our yard so we ate the ackees.

The ackee and breadfruit are famous because Royal Navy Captain William Bligh took the breadfruit to the Caribbean as slave food. He also took the ackee from Jamaica to the Royal Botanic Gardens in Kew 1793. The Jamaican national dish is ackee and saltfish, and that saltfish is cod, which was sent from Scotland to the Caribbean to feed the slaves. So, the two trees in my little backyard were part of the very important story of British history . . . the navy and slavery. Those two trees are iconic.

By 1955 my mother had saved enough money to bring me and my brother over to London. One of the things I noticed when I arrived here was that you had trees growing out of concrete in the road. After I arrived in London, I thought, Why are trees growing out of the concrete? But in the summer, they were nice and it changed the atmosphere of the place, which was sometimes pretty bleak. We were living in a very tough part of Highbury, London.

I arrived in London when I was fourteen years and eleven months old, one month short of the school-leaving age in 1955. I had hoped not to go to school at all but the authorities checked, and insisted that I attend school. I went, hoping to leave in one month at Easter but the headmaster refused to take me for just a month. He said that I had to stay till the summer term.

Not long after that everything changed because I was very good at cricket. This appeared in the local paper and the headmaster of the local grammar school insisted I was transferred, because he needed a cricketer.

I struggled in the grammar school, but one of the subjects I found easiest to cope with was Botany. This was because we had a teacher called Mr Youngs, who was the biology teacher, who took an interest in my education.

Trees produce an ingredient – oxygen – which we cannot see or smell but is vital for our survival. We have a relationship with trees and we will never work out how that relationship developed. That is one of the mysteries of life. When people talked about hurting animals, I used to say, "How do you know trees don't want to be hurt? They're living. You've got to treat them with respect because they're alive."

I worked with barley and would not throw away a grain unless I had to. When I think of the importance of trees and plants, I think of the song, "John Barleycorn", which is famous because of Robert Burns.

It's a song about planting a small grain, apparently insignificant, then it grows and then, how, in order to destroy it you cut it down, at the knee, then it is knocked around, threshed, soaked in water, grows, is kilned. The poem ends by saying:

"Then let us toast John Barleycorn;
Each man a glass in hand;
And may his prosperity
Never fail in Old Scotland!"

An apparently insignificant thing, a grain, has this great potential. Prejudgement is wrong. That is also the case with plants – a mango tree, a breadfruit tree or an ackee tree – they all have some important links with our survival and our culture and you cannot begin to speak about the importance of wheat. In terms of black history you have to talk about sugar cane, coffee, cotton and tobacco!

TONY KIRKHAM
Head of Arboretum, Kew

One of my favourite trees is the Douglas fir. I recently went out to Vancouver Island, to Port Renfrew, to meet Big Lonely Doug.

What a great meeting. You're in the middle of a block of old growth forest that was cleared in 2011 and Doug, this giant Douglas fir, is still in there, where they had left him, on his own. It was Dennis Cronin, the forest engineer who was marking up and valuing the timber on that block, who came across it and decided he couldn't fell it. He taped it to say "leave tree". He just thought, oh this is an amazing specimen, and actually it would have been a really valuable timber tree. It was the second tallest tree in the country, and is around seventy metres tall. They cleared the whole rest of the forest, about twelve hectares, and there's Doug, standing on his own.

BONNIE FAIRBRASS
Comedian and activist

I can remember as a child seeing photos of the great redwood forest in America and being really angry about them being cut down or having pathways cut through them. It physically hurt me.

The onset of climate change has been the same for me. I feel an overwhelming urge to protect all of the spoils of mother nature, but trees especially so. I am basically the Lorax, with less facial hair, thankfully.

SANDY BENNETT-HABER
Writer

If I go out our front door and cross the road I can get a bus to Edinburgh's Princes Street in ten minutes, yet from the big window at my kitchen sink I look out on to a forest. The scene is dominated by a row of cypress; once upon a time they were planted as a hedge to subdivide our communal garden, now they are so tall you have to lean your head back to see the treetops.

Legend has it that on the day I was born, while my parents were living near the beach in a tent on the south-east coast of Australia, my father bought a big patch of land on the edge of a nearby hamlet called Cabbage Tree Creek. Eucalyptus, tree ferns, dogwood and acacias were my first playground

Our city life in Scotland is a long way away from that temperate Australian bush. When we first moved in, the forest was a tip. Years of fallen branches, garden clippings, dumped building materials and rusted wire made the treeline dark, inaccessible and uninviting. I would try to imagine the space transformed – but the job seemed insurmountable. What impact could I possibly have? I had tiny children, a husband who worked away, my family was on the other side of the world and I was desperate to reconnect with my own writing. Time amid the rot just reminded me of the impossibility of ever getting anything done.

The seasons changed, my children grew and, despite the city on our doorstep, I often found myself called to our garden, pottering among the branches and leaf litter, clippers in hand.

Today the spider and wire-infested tangle has diminished. My young self, striding confidently through the trees is delighted to see my children climb, jump, dig, picnic, build and explore under the dark green canopy of our forest. As I worked twig by twig to reclaim the forest floor I came to see the difference small actions can make. Could I plant seeds? Improve the soil? Could I buy less plastic and have an impact on climate change?

The leaves of the conifers whisper with their neighbours, this high-up conversation indifferent to the humans far below. But from my kitchen sink, or standing looking up through the leaves, I no longer feel insignificant. I see how small actions gathered together over time can change the world.

TREES LIKE US

What has blown me away, as I wrote *For the Love of Trees*, was the way people I spoke to put into words ideas, thoughts, emotions I had only half-grasped or felt. They told me stories of distant forests, bringing them to life, and they spun tales that made me imagine I was there with them, watching raindrops bounce under a holly tree umbrella or leaning up against a pine and feeling it move in the wind.

They taught me that the ways in which we connect with trees are many. There are the physical ones, but there are also the psychological ones. We see ourselves in trees, or trees in ourselves. We can see the veins of a leaf beneath our own skin, the struggle against adversity in a tree bent over.

STEFAN BATORIJS
Nature therapist

I've had this strong connection with nature for all of my life, since I was able to walk. An encounter with a pine tree was the catalyst for the spiritual experience of shinrin-yoku.

I was at a turning point in my life and I'd managed to get on to a week-long residential course at Schumacher College. It was in April and it was that strange April weather where it can be really hot and then suddenly you get these big showers with thunderstorms. We'd had that all week, so it was quite dramatic, and I remember one evening it was raining, but it was that warm wet rain, and we were sent off on our own to go off into nature and meditate. It was twilight, a dusky light – the gloaming. I walked past this gate into an experimental forest garden; as I walked past there was a bit of me that couldn't resist going in. It was beckoning to me. There was just the sound of the rain, stillness. I climbed over the gate and I wandered around entranced by this beautiful forest garden, in this lovely wet rain.

Then I saw this pine tree, small and stunted, almost like a bonsai, and so I climbed in underneath it to shelter from the rain and I nuzzled my back up against its trunk. I sat there for a while, listening to the rain, and then I looked up into the branches, which were only just above my head and there was a tiny handmade doll, about four inches high. The doll had hair and clothes on and it looked identical to my partner at the time – same hair, same clothes. I started to go into a kind of altered state. I was looking at the doll and thinking, *what does that mean?* As I was thinking, I was hit by this really strong scent of the tree. Suddenly the pine scent was so strong and profound. I was just breathing it in.

Then it occurred to me that the pine was going inside my body, into my blood stream. I thought, *Wow.* The tree is inside me. I was blown away by that and I sat with that for quite a while in the rain, thinking, *The tree is inside me.* Just sat with it and then after a while I recognised that as I was breathing out, the tree was breathing me in as well, so there was part of me inside the tree. The tree was taking the air as part of its photosynthesis process. So I felt that there was this reciprocity between the two of us, that suddenly we were in a cycle, linked together. As I breathed in the tree was inside me; as I breathed out, I was inside the tree.

PRATYUSHA
Poet

One of the things I love about trees is that they mirror a sense of interconnectedness – their leaves reflect the veins on the back of my hand, or the crossings in a map or rivers. Thinking of that makes me feel close to them. You can never feel alone around a tree.

One of my favourite trees is the banyan tree. It has aerial roots that plunge from the branches right down to the ground and then take root, resembling trunks. There's a lovely poem by Craig Santos Perez. He writes about how these branches/roots/veins form their own pathway of connecting to the air and the ground and water. It's a network or a bridge across air.

It's something I really enjoyed looking at as a child, because I used to find it so magical that a tree could grow like that. The banyan draws curiosity from a child, particularly the tactile quality of those roots. When I lived in Bombay, there was a large banyan near my apartment. It became a school bus stop. Everything grew by/ in the banyan – birds in its branches, fruit and flower offerings by its trunk, insects crawling over its roots, children waiting under its large shadow in the tropical heat. There's an entire ecosystem embedded within the tree. It is an organism growing over all of us, its roots entangled in memory.

KATHERINE MAY
Author

The Blean on the outskirts of Canterbury is such an important place to me. I often used to go and walk there to decompress after work, and it was the starting point for *The Electricity of Every Living Thing: One Woman's Walk With Asperger's.*

Walking there one afternoon, I stopped paying attention to the waymarkers and found myself lost deep in the forest with no mobile phone signal or map. It should have been frightening, but it was exhilarating instead. I wandered for three hours and got thirsty and exhausted before I found my way back to my car again. A couple of hours in, I stopped for a break and realised the whole wood was crackling with life – I could hear it, the drawing up of water, the sense of everything growing incrementally around me, and the feeling of life concealed everywhere within it. It was spiritual moment that changed me profoundly.

When my son stopped coping in school and we had to pull him out, the first thing we did was go to the woods and build dens. It was midwinter, and freezing cold, but we brought flasks of tomato soup and chocolate, and it felt good to make a safe place for ourselves. There's one den that's still standing in the Blean, which we built in those first few weeks when everything was falling apart. We now visit it and see that other people have repaired it and added extra sticks. I like to imagine that other children are finding comfort in it now – we've passed on a way of surviving.

I think trees serve as metaphors for humanity – we project a lot on to them. When they're young, they're skinny and fragile, but also surprisingly resilient. As they age, they gain in stature and become gnarled with wisdom.

ALL BRANCHES ARE ONE

The poet Jackie Kay told me, "We're all linked and we're not just linked through the sea – we're linked through trees. Trees control so much of everything that goes on and people are ignorant to their wisdom."

Everything comes back to that idea – that we are all linked. Often, we separate stories out. We talk about the science of how trees make us less stressed, of how forest bathing can help our mental health. Or we will read news stories about how many trees we need to plant to save the planet. But these are not separate stories. They link, entangle and fuse. They are themes in the same story of how we and the environment, the natural world, are intertwined, and one.

15

WHAT CAN WE DO?

HOW TO FOREST THE FUTURE

The best time to plant a tree was twenty years ago. The second best time is now.

CHINESE PROVERB

So, what can we do? If trees matter so much – to humans as individuals and as a species, and to the wider planet, with or without us – how can we help them spread and thrive?

We've put together a list here of things we believe make a difference. Some of these acts are small – too small, perhaps, to register if done alone, but meaningful when enough of us do them. They are like tiny acorns, rich with potential. We don't know whether they will grow to become big, mighty oaks, whole forests, or barely make it out of the ground.

PLANT TREES, WITH THOUGHT AND CARE

The UK Committee on Climate Change has advised that we need to plant 90–120 million trees each year to hit our 2050 net zero targets. So we must plant – while also remembering that, when it comes to emissions, planting is not enough and that we can't just wash ourselves green by planting trees.

But once you start listening properly to stories about trees, you realise that change isn't as simple as plonking millions of saplings in the ground. It's about the right tree, in the right place. Figures released by the Forestry Commission in 2020 showed that even in the midst of our fervour for tree planting, the area of woodland in England that is managed sustainably has dropped – by 10 per cent since 2015.

So, how to choose the right tree and the right places? There are plenty of resources to help with this, from Max Adams *Little Book of Planting Trees* to the websites of the Woodland Trust and the Royal Horticultural Society.

SIGN THE CHARTER FOR TREES, WOODS AND PEOPLE

The words of this document, written by author Fiona Stafford, are a poetry that incites action. "Natural treasures," it begins, "in roots, woods and leaves, for use, the air that we breathe. Imagine: a wood starts with one seed. We're stronger together – people and trees." The charter, a document that came out of a Woodland Trust initiative to create a unifying statement about the rights of people in the UK to the benefits of trees, woods and forests, is also a real community statement. It brings together the ideas of countless people across the UK including three hundred community groups. Absorb it, sign it, help start the journey we need to do together.

GROW A TREE FROM SEED

There are few things more magical than growing any kind of plant from seed – even better a tree. So, if you have time and a place for your tree to go, why not start it off yourself. Take your apple pip, beechmast, hazelnut, acorn, or whatever it is you have and put it in the ground or compost and watch the miracle unfold. The Woodland Trust website has a simple set of instructions for how to do this and also a DIY tree protection video if you're worried about nibbling rabbits.

REWILD YOURSELF

Let yourself go feral. Take a wander into the deep, dark woods, allow the moss and the bracken to stick to your clothes. Climb a tree, but make sure it's not an ancient tree or one whose branches are likely to crack and splinter. Sleep out with nothing but the canopy between you and the sky. Build a den. Make a shelter from fallen branches and leaves. Put up a tent in the forest. Let the wild woods take hold of your heart. But make sure you have permission from the landowner to wild camp if you are in England or Wales, and, wherever you are, do it responsibly.

HELP OTHERS PLANT OR LOOK AFTER TREES

Whether you're putting a few trees in the ground or a whole woodland, get other people involved.

Tree planting done together is so much better than when it's done alone.

PLANT OR GET INVOLVED IN A LOCAL COMMUNITY ORCHARD

The community orchard movement is one of the most glorious, life-affirming ways of bringing us all together around trees. Plant an orchard and you can harvest apples, press them, make cider, and sing your lungs out on a wassailing session.

BECOME A FOREST SCHOOLS LEADER

It's all about passing on the love, isn't it? And what better way of doing it than take a bunch of kids out into the woods to play, connect and get to know the trees.

GO ON A TREE-PLANTING BREAK

As with all planting, you want to make sure it's the right trees in the right places with the right organisation. Trees for Life, for instance, offer a conservation week in the Scottish Highlands for volunteers that involves "growing trees, planting trees, monitoring wildlife" in the Caledonian Forest.

PLANT YOURSELF IN A SIT SPOT

When we walk noisily through the forest the wildlife slinks away and hides. A cone of silence envelops us and we do not hear the forest speaking. But when we choose a spot and sit for a while, then the life of the woods starts to emerge, and we become alert to its presence.

BE A TREE DEFENDER

Protecting what we already have is as important as reforesting and rewilding. Only 2.4 per cent of the UK is covered by ancient woodland. These are trees that are over four hundred years old in England and Wales, or over 250 years old in Scotland. Figures released by the Woodland Trust in early 2020 showed that they were aware of 1,064 ancient woodlands that were at risk of damage or destruction – the highest number since the Trust started compiling the data in 1999.

Join the Ancient Tree Forum or sign one of the petitions such as Janis Fry's, a call to Save Britain's Ancient Yews, or the petition for Legal Rights For Ancient Trees.

HUNT DOWN THE ANCIENTS

The Ancient Tree Forum offers a number of

ways to help protect ancient trees – perhaps most thrilling being by volunteering as a recorder of ancient and veteran trees. Better still, call yourself an Ancient Tree Hunter.

ORDER, ORDER

One way of defending trees is to set up tree protection orders for those we think might be in danger, or even for trees we simply value. Friends of the Earth has a clear set of instructions for how to do this on their site.

SIT IN A TREE OR SUPPORT A PROTEST

The battle to save trees happens on multiple fronts. The current HS2 protest shows how many people are needed in different ways. You can fight by camping at threatened woods. You can do it by putting up a legal challenge, as Chris Packham has.

But you can also do it in other smaller ways – by signing petitions, writing to MSPs, donating money and making a noise on social media.

LEARN YOUR TREES

If you don't already know what the trees on your local street, or in your local park are, find out. Pick up a book or download an app like Tree ID from the Woodland Trust. Then, once you do know those tree names, you can always spread the word by joining the graffiti botanist craze which swept the UK during lockdown and chalking them across the pavement to provide inspiration for others.

KNIVES OUT

There is a spoon in every fallen branch and a wand in every twig. The UK, according to

London's most famous whittler, Barn the Spoon, is going through "a woodculture renaissance". Whittling isn't difficult to master and guides to how to do it can be found online and in books – for instance, Barn's *Woodcraft*, which will tell you all you need to know about green woodworking.

USE SPARINGLY, WITH LOVE

What we consume has an impact on what kind of forests exist out there, globally. If we want to have diverse, mature forests across the globe, we need to think about a whole range of our different consumptions – including the foods we eat, like palm oil, soy and cocoa, which might be grown on plantations that are replacing forests. Ask yourself is that beef on your plate soy-fed?

We also might want to look at our consumption of tree-based products, like paper and packaging. Out of the 17 billion cubic feet of trees deforested each year, over 60 per cent are used to make paper.

TELL YOUR TREE STORY

Shout it, sing it, draw it, photograph it, and send it out on social media, ideally tagging in @treestoriesuk so we can spot it. The more we share our pleasure, the more we show the world how much trees count. The Charter for Trees, Woods and People has a section urging us to "celebrate the power of trees to inspire". As its author Fiona Stafford says, "Stories have always grown on trees. Artists are drawn to their intricacies. Woods are rooted in memories, but it's the leaf mould of tales told that nourishes future growth." Aye, aye.

REMEMBER YOUR GREEN FOREST CODE  Forestry England

When we visit the forest it's important we treat it, and others there with respect. One of the best guides to how to do this is Forestry England's Forest Code

PROTECT AND RESPECT WILDLIFE, PLANTS AND TREES
When spending time in the forest, consider your surroundings and make sure that all plants and animals are left undisturbed.

GUARD AGAINST ALL RISKS OF FIRE
Dry weather makes it very easy for wildfires to start. Help us prevent this by not lighting campfires, only use BBQs where allowed and make sure any cigarettes are put out and disposed of properly. If you see a wildfire dial 999.

KEEP DOGS UNDER CONTROL
We know the forest is an exciting place for any canine companion! That's why it's important to make sure your dog is under control at all times, for your dog's safety, yourself and the safety of other visitors. Read the full Forest Dog Code here.

TAKE YOUR LITTER HOME
Don't be a rubbish visitor! Litter is unpleasant for visitors but can also cause damage to the landscape and wildlife. Most of our visitors take their litter with them, please do the same.

MAKE NO UNNECESSARY NOISE
Forests are a great place to escape the hustle and bustle of everyday life. Avoid disturbing wildlife and other visitors by keeping noise to a minimum. Ditch the loud music, any unnecessary shouting and revving of car engines and just enjoy the peace and quiet.

TAKE ONLY MEMORIES AWAY
Anything that belongs in the forest should stay in the forest. Respect the forest and the wildlife that lives there by leaving their home intact.

THE FOREST IS FOR EVERYONE. PLEASE BE AWARE OF OTHER VISITORS
From young families to adventurous bike riders and dog walkers to bird watchers; we want everyone to enjoy their visit to our beautiful forests. Be considerate of other visitors when moving around the forest. Why not offer a smile and friendly greeting? Everyone should feel welcome!

THANK YOU

Thanks to the tree lovers who made this book possible by giving us their stories. You taught us new ways of relating to trees.

Thanks to our parents for making us tree climbers and stick collectors and lovers of bark and lichen. Thanks to our husbands, Andy and Rob, and our kids, Louis, Max, Lily, Finlay, for putting up with another mad project in a particularly mad year – but also being the ones we most love to hide in the woods with.

Thanks to our agent Jenny Brown and to the team at Black & White, who have given our book such care and love during difficult times.

Thanks to Karen Glossop and the Woodland Trust's Tree Of The Year for inspiration. But most of all thank you to those whose names we do not know who planted the trees in our parks and on our streets that make our lives better.

And, of course, thanks to the trees themselves. We owe you everything.

IMAGE CREDITS & PERMISSIONS

All photography except images on the pages listed below copyright © Anna Deacon 2020

Headshots on pages 9, 13, 16, 17, 22, 29, 31, 32, 41, 42, 46, 60, 74, 75, 76, 80, 81, 91, 98, 106, 109(bottom), 112, 116, 124, 125(left), 127, 131, 132, 135, 144, 148, 152, 163, 173, 180, 181, 186, 187(bottom), 201, 202, 203, 208, 217, 222, 225, 234, 237, 238, 239 courtesy of the contributors

8: Craig Stennett/Alamy Stock Photo
11: Simon J. Evans   **14:** Paul Wilkinson
16: Kiliii Yüyan   **28:** Alisa Connan
44, 45: Joanne Crawford
78, top & 79: Julian Eales/Alamy Stock Photo
90: Iain Turnbull   **97:** Rob Deacon
104, 105: Talia Woodin
107: Redorbital Photography/Alamy Stock Photo
114, 115: Hana Wolf, studio-wolf.co
125, left: Martin Shields
125, right: Christopher Jones/Alamy Stock Photo
132: Fiona Higgins   **138:** Matilda Temperley
144: Liz Brazier

Permissions

iii: Alycia Pirmohamed, 'My Body Is A Forest', published in the *Adroit Journal*, reproduced by kind permission of the author.
7: Jackie Kay, 'The World of Trees', reproduced by kind permission of the author.
51: Carol Ann Duffy, 'Forest', reproduced by kind permission of the author's agent, Peter Straus.
65: Sean Wai Keung, 'Where is the tree my gongong drew?', first published in bathmagg Issue 4, reproduced by kind permission of the author.
87: Kathleen Jamie, 'The Wishing Tree', reproduced by kind permission of the author's agent, Jenny Brown.
125: India Knight, edited extract from *Sunday Times* article, 16 February 2020, © The Sunday Times / News Licensing.
177: Pratyusha, 'if still forest (winter)', reproduced by kind permission of the author.

FURTHER READING

Max Adams, *The Wisdom of Trees*, Head of Zeus (2014)

Barn the Spoon, *Spon: A Guide to Spoon Carving and the New Wood Culture,* Virgin Books (2017)

Adrian Cooper (editor), *Arboreal: A Collection of New Woodland Writing*, Little Toller Books (2016)

Roger Deakin, *Wildwood: A Journey Through Trees*, Penguin (2008)

John Evelyn, *Sylva: Or a Discourse of Forest-Trees, and the Propagation of Timber in His Majesties Dominions*, Forgotten Books (2018)

Jean Giono, *The Man Who Planted Trees*, Vintage (2003)

Jay Griffiths, *Wild: An Elemental Journey*, Penguin (2014)

Isabel Hardman, *The Natural Health Service: What the Great Outdoors Can Do for Your Mind*, Atlantic Books (2020)

Jim Hindle, *Nine Miles: Two Winters of Anti-Road Protest*, Underhill (2016)

Sarah Ivens, *Forest Therapy: Seasonal Ways to Embrace Nature for a Happier You*, Piatkus (2018)

Kathleen Jamie, *Surfacing*, Sort of books (2019)

Jackie Kay, *Fiere*, Picador (2011)

Eduardo Kohn, *How Forests Think: Towards an Anthropology Beyond the Human*, University of California Press (2013)

Dara McAnulty, *Diary of a Young Naturalist*, Little Toller (2020)

Wangari Maathai, *Unbowed*, Arrow (2008)

Sara Maitland, *Gossip from the Forest: The Tangled Roots of Our Forests and Fairytales*, Granta (2013)

Joan Maloof, T*eaching the Trees: Lessons From The Forest*, University of Georgia Press (2010)

George Monbiot, *Feral: Rewilding The Land, Sea and Human Life*, Penguin (2014)

Beth Moon, *Ancient Trees: Portraits Of Time*, Abbeville Press (2020)

Thomas Pakenham, *Meetings With Remarkable Trees*, W&N (2015)

Robert Penn, *The Man Who Made Things out of Trees*, Penguin (2016)

Richard Powers, *The Overstory*, Vintage (2019)

Jim Robbins, *The Man Who Planted Trees: The Story of Lost Groves, The Science of Trees and a Plan to Save the Planet*, Random House (2015)

Alice Roberts, *Tamed: Ten Species That Changed Our World*, Windmill Books (2017)

Shel Silverstein, *The Giving Tree*, Particular Books (2010)

Fiona Stafford, *The Long, Long Life of Trees*, Yale University Press (2017)

Paul Sterry, *Collins Complete Guide to British Trees*, Collins (2008)

The Good Seed Guide: All You Need To Know About Growing Trees From Seed, the Tree Council (2001)

Isabella Tree, *Wilding: The Return of Nature to a British Farm*, Picador (2018)

Luke Turner, *Out of the Woods*, W&N (2019)

Peter Wohlleben, *The Hidden Life of Trees*, William Collins (2017)

Paul Wood, *London Is a Forest*, Quadrille (2019)

Holly Worton, *If Trees Could Talk: Life Lessons From the Wisdom of the Woods*, Tribal Publishing (2019)